WARMING UP

WARMING UP

How Climate Change is Changing Sports

Madeleine Orr

BLOOMSBURY SIGMA
LONDON · OXFORD · NEW YORK · NEW DELHI · SYDNEY

BLOOMSBURY SIGMA
Bloomsbury Publishing Plc
50 Bedford Square, London, WC1B 3DP, UK
29 Earlsfort Terrace, Dublin 2, Ireland

BLOOMSBURY, BLOOMSBURY SIGMA and the Bloomsbury Sigma logo are
trademarks of Bloomsbury Publishing Plc

First published in the United Kingdom in 2024

A catalogue record for this book is available from the British Library

Library of Congress Cataloguing-in-Publication data has been applied for

ISBN: HB: 978-1-39940-4-525; eBook: 978-1-39940-4-518

2 4 6 8 10 9 7 5 3 1

Typeset by Deanta Global Publishing Services, Chennai, India
Printed and bound in Great Britain by CPI Group (UK) Ltd, Croydon CR0 4YY

To find out more about our authors and books visit www.bloomsbury.com and sign
up for our newsletters

For Nico, Ari and Dalia.

Contents

Preface

In the fall of 2015, I first noticed that climate change is changing sports.

On a gap year, I traveled to the French Alps to work as an Overseas Operations Coordinator for a small British ski company. It was a seasonal job, and I didn't expect it to be particularly challenging; I would be responsible for booking bus transfers between airports and many resorts across the French and Italian Alps. I'd also be responsible for the add-ons of each group's trip, like ski rentals, lessons, lift tickets, and the occasional round of pub trivia.

At the resort, staff were housed four to a room on bunk beds, in the hotel's basement. Not exactly glamorous. But it was straightforward work, made better by great people.

The ski conditions, however, left much to be desired. Originally, an early dusting of snow in November left us hoping the mountain would open the first week of December. But the snow gods weren't operating on our schedule. Opening day got pushed back to mid-December, and was only possible thanks to snowmaking technology, with limited open runs. These conditions remained until the second week of January. I remember because the general vibe in town for that first month reflected the poor conditions on the mountain: glum and grey. My dad and stepmom visited for a week at Christmas, but skied only once because conditions were so sub-par. They returned their skis mid-week.

In the four weeks between when the season was supposed to start and when the real snow finally arrived, the tourism industry suffered. Several small restaurants in town changed their schedules to open for fewer days each week or cut their lunch service to save on labor costs. Shopkeepers at the three separate ski-rental stores near our hotel told me their sales were down compared to the year before because tourists were cutting their ski holidays short, or not coming at all. And it wasn't just an economic problem, though that alone could be catastrophic to a place like this.

Early-season injury rates were unusually high. On bad snow, slush or ice, the rate of accidents – particularly among unskilled skiers – seems to increase. Adding insult to injury, when conditions don't allow for all the runs to be opened, ski areas will typically funnel skiers onto a few runs where snow guns can ensure reliable conditions. The extra traffic on these runs increases the chance of collision, especially among inexperienced skiers and children. As part of my job with the ski company, I would occasionally be sent to the hospital to check on an injured client in the emergency room. Every time I went, the waiting room was full. Broken wrists and dislocated shoulders for snowboarders, knee injuries for skiers.

On my third visit, I asked the triage nurse whether this was normal. She told me injury rates were high. The radiologist had a similar answer to the same question: he likened the injury rates on this early January Tuesday to "the busiest Saturday when all the kids are here".

When the heavy snow finally arrived, local skiers got excited and took more risks than usual. I know because I was one of them. Three days after the first big snowfall, I went off-piste skiing and caught the tip of my ski on a section of ice under the powder, which destabilized me, keeping my right leg locked in place while the other barreled full speed ahead. I bailed and landed on rocks, damaging two of the ligaments in my knee and obliterating both menisci. Under my snow pants, I could feel the swelling stretching the fabric of my leggings. By the time I got to the bottom of the hill and took my snow pants off, my knee had turned purple. I knew my season was over.

Because I was on a work visa that would expire at the end of the ski season, and the surgeries and recovery would take much longer than that, the doctors decided I should be repatriated to Canada. But first, I would have to stay in France for three more weeks to stabilize the injury and ensure I wouldn't develop a blood clot on the high-altitude flight home. In the weeks between my injury and my repatriation, my life changed.

Each day, I sat on a single bed in a small hotel room with salmon-colored walls. I hated those walls. I was grateful that the hotel staff put me in a guest room so I'd be on a floor with an elevator. I spent most of my time watching the news, but to break up the mundanity, I'd take shaky walks on my crutches through the town. Because I was moving about as slowly as a two-year-old, I noticed the subtle signs that something was *still* off, even though the snow was back.

Then, on the evening news, which I dutifully watched from the hotel bar each day, stories of avalanche deaths started popping up. In one case, a group of teenagers were killed in an avalanche at Les Deux Alpes, one of our neighboring mountains. A few weeks later, several Czech skiers were killed by an avalanche while on holiday in Austria. Sure, avalanches happen. But the first half of the 2016 season was especially deadly.* In town, I noticed avalanche warning signs and educational materials displayed in the window of the ski ticket office.

---

*A few years later, the 2020/21 ski season would become the deadliest on both sides of the Atlantic since the "Winter of Terror" – a three-month stretch in the 1950/51 ski season that saw a series of unprecedented avalanches along the Swiss-Austrian border. An estimated 650 avalanches fell in 1950/51, killing more than 265 people and causing significant damage to several ski villages. Of course, in 2020/21, COVID restrictions led many resorts to open only partially, sending many avid skiers to backcountry trails where risks are higher.

On one of those days, I took a cab to the neighboring town for lunch with a colleague. We ate in a building that must have been hundreds of years old, at a little stone-walled restaurant. At the table next to us, a group of older men were chatting in frustrated tones about the possibility of selling their condos on the mountain to buy at a resort with more reliable snowfall. These men worked as ski instructors on the mountain; their logoed jackets gave them away. This is a not-uncommon feature of the resorts in the French Alps, where retirees will live in the mountains for the winter season to teach ski lessons a couple of days a week in exchange for a season pass and some extra income. These men had been skiing in the region their whole lives. The waitress knew their names. This whole community lived and died on the ski season. And they were ready to leave.

In those weeks of hobbling around town and working from my hotel bed, I got an acceptance letter via email to the PhD program in Kinesiology at the University of Minnesota. The acceptance provided exactly the motivational boost I needed to think more critically about what I was seeing out of my window in the Alps.

The weight of the full-length knee brace I wore served as a constant reminder that the economic and health impacts of a bad snow year were deeply consequential for ski communities. I'd grown up skiing at small resorts in eastern Canada, but I'd only ever been there as a tourist. Back then, a bad snow year was simply an inconvenience, a bit of bad luck. Now I had met some of the people whose livelihoods suffered when the snow wasn't good. It felt personal. It only took a few minutes of searching online to confirm what I already suspected: this was climate change.

The consequences of climate change for the sports sector became clear to me during my season in France. And that got me thinking: it can't just be skiing.

Four months later, with the first of several surgeries behind me and a break before starting my PhD program, I was looking for something to do, and landed two jobs that would bring me to the 2016 Olympics. Both jobs were essentially the same thing:

supporting the moms, dads, partners and siblings of athletes competing in the Games.

And that's how I found myself in Rio de Janeiro for the summer of 2016. It was incredible, and jarring. Rio is a stunning city, juxtaposing fancy high rises in the beachside neighborhoods of Leblon, Ipanema, and Copacabana, with favelas that sprawled up the hillsides nearby. The city has the best açai and sorbet I've ever tasted, and phenomenal panoramic views from the summit of Corcovado. But it's also very dirty.

Rio was already overwhelmed by waste: the year before the Games, 32 tonnes of dead fish were pulled from the Rodrigo de Freitas lagoon where canoeing and rowing events were due to be held. Officials cited pollution and water temperature as the causes. Guanabara Bay, where sailing, windsurfing, and open-water swimming races were hosted, was also highly polluted. In 2015, during a sailing competition in the Bay, organizers reported that 25 per cent of athletes were affected by nausea and diarrhoea. Brazilian authorities promised a 4-billion-dollar clean-up in and around the bay, but this never materialized. Only 170 million was committed to the cause between 2015 and 2016 and conditions were not much improved. The Olympic organizers created plans to test the water at each venue on a regular basis, and did their best given the time crunch and available resources. In the end, athletes were told to swim with their mouths closed.

In Rio, I had another rude awakening. The number of people, and thus the amount of consumption, at these events is immense. More than 500,000 people traveled to Rio for the Games, each consuming goods, using energy, and emitting pollution through their travels. You can imagine the consequences. Waste systems were overrun by the 17,000 tonnes of waste produced by the event, which only added to the city's regular high waste stream.

Organizers estimated the venues used 29,500 megawatts of electricity. Official vehicles, which included 1,500 buses for athletes, staff, and volunteers, consumed 23.5 million litres of fuel. In total, it is estimated the Games emitted 3.6 million tonnes of

CO_2. Even excluding fan travel, the event produced more carbon than 26 countries with a combined population of more than 10 million over the whole of 2016.

Sport's carbon footprint and contribution to waste is massive. In the current climate, it's irresponsible to ignore this.

From the beaches of California to the islands of the South Pacific, from the Alps to the Horn of Africa, climate change is threatening sport in every corner of the world. The realities of climate change are now undeniable and likely to worsen. And sports – especially outdoor sports – are on the frontlines; athletes are outside day in and day out, witnessing these changes. They're seeing climate change in real time, and it's taking a toll on everything from scheduling to on-field performance, from mental health to the financial bottom line.

The question isn't whether climate change will impact sports. It already is. The question is: how fast can the sports world adapt?

CHAPTER ONE
PREGAME

In 2020, amid the COVID-19 pandemic, we got a glimpse of life without sports. In a slow domino effect originating in China, then flowing through Europe and North America, the sports industry ground to a halt.

During this pause, for those of us working inside or adjacent to the sports industry, we saw sports adapt. While youth sports and physical activity for people of all ages is essential, with health and social benefits that have been well established through decades of research, it became clear rather quickly that sports as a form of entertainment at the pro and elite levels were not so essential. Sports infrastructure, on the other hand, could be.

A few days after the professional sports sector shut down, organizations and athletes stepped up, including tennis icon Rafael Nadal, basketball stars Zion Williamson and Giannis Antetokounmpo, and American football player Drew Brees. Each player made personal donations to support those impacted by the virus through loss of employment, food shortages, or industry disruptions. More than 100 athletes from 20 sports donated memorabilia to the "Athletes for COVID-19 Relief" fund, which were raffled off to benefit the Center for Disaster Philanthropy's COVID-19 Response Fund.

By the end of March 2020, sports edged closer to the frontline of the pandemic response. Fields and stadiums in every continent were converted into temporary hospitals, shelters, and emergency resource distribution centers. Very few of the makeshift hospitals actually saw patients; the anticipated patient flow was greater than the actual number of hospitalized COVID patients. While not purpose-built for emergency scenarios, sports facilities make a lot of sense as emergency response sites due to their size, amenities, and durable design.

The COVID pandemic was not the first time sports facilities had been used for emergency services: there is a decades-long history of using stadiums, gyms, arenas, and community centers to house people during natural disasters (I'll talk about this more in Chapter 8). The difference here was that the COVID pandemic was acute – global in scale – and variations of the same emergency shelter plan were popping up in cities across the globe.

As Greta Thunberg, David Attenborough, and Al Gore have been warning for years (some longer than others), there's another global crisis underway. According to the United Nations, "climate change is the defining crisis of our time and we are at a defining moment". In 2019, David Carrington, Environment Editor at the *Guardian*, penned an update to its internal guidelines for language on climate change: "Instead of 'climate change' the preferred terms are 'climate emergency, crisis or breakdown' and 'global heating' is favoured over 'global warming', although the original terms are not banned."

These terms reflect a growing movement at the city, state, and national level in jurisdictions around the world to declare climate emergencies. As I write this, 1,986 jurisdictions, including 15 nations, and covering more than 1 billion citizens, have declared a climate emergency.* Unlike COVID, the climate crisis is a slow burn, one that can be hard to pin down and easy to pass off as someone else's problem. And yet, the climate crisis is just as global in scale as COVID, will take lives, and stands to slow or even halt the economy in some industries or regions when the worst of

---

*The Climate Emergency Declaration website tracks climate declarations. It is updated regularly. https://climateemergencydeclaration.org/

its impacts are realized. The biggest advantage we have with the climate crisis is advance notice.

How did we get here?

When I was born, there were 357.85 parts per million (ppm) of carbon in the atmosphere. As I write this, not 30 years later, there are 419.96ppm. This is the greenhouse gas effect in action: the Earth's surface is warmed by long-lasting gases trapped in the atmosphere, mainly carbon dioxide (CO_2), nitrous oxide (N2O), and methane (CH4), alongside a range of industrial gases. These gases allow sunlight to pass through and warm the Earth's surface, but prevent some of the heat from escaping back into space, causing the planet to warm up. This natural phenomenon was first described by French mathematician Joseph Fourier in 1824. Yes, that's 200 years ago.

Over time, several warnings about the risks of runaway emissions have been issued to world leaders. One of the earliest warnings came in 1965, when a report by President Lyndon B. Johnson's Science Advisory Committee warned of the potential consequences of increasing CO_2 levels in the atmosphere. Seven years later, in June 1972, the UN Scientific Conference, also known as the First Earth Summit, was held in Stockholm, Sweden, where a declaration was adopted for coordinated international environmental action. In a section on the identification and control of pollutants of broad international significance, the declaration raised the issue of climate change for the first time, warning governments to be mindful of activities that could lead to climate change and evaluate the likelihood and magnitude of climatic effects.

Over the next 20 years, concern for atmosphere and environmental issues grew. In the 1960s and 1970s, the US, Canada, the USSR, India, and Europe saw the fastest rollout of environmental laws of any period in history. That was also the era that gave us Earth Day, the World Wildlife Fund, the Environmental Protection Agency in the US, the Canadian Wildlife Federation, Greenpeace, and the world's first international environmental convention: the Nordic Environmental Protection Convention of 1974. The environment was popular, and it's easy to see why. Find me one parent who would prefer for their kid to grow up without clean air, clean water, and a nice place to play outdoors. I'll wait.

However, in this same period and through the 1980s, another project was afoot across most of the Western world: neoliberalism, aka widespread deregulation and free-market policies to promote capitalism. Economic growth, measured in GDP, was the dominant metric for success. This is still true in the 2020s. Environmental regulatory power was taken out of government hands and put squarely in the corporate sector. The era of voluntary action was born.

In 1992, the international policy community met in Rio de Janeiro for what was called the Rio Summit. This conference birthed the United Nations Framework Convention on Climate Change, the UN agreement ratified by 198 countries to prevent dangerous human interference with the climate. Three years later, in 1995, the parties to the framework (representatives from signatory countries) met in Berlin for the first of several Conferences of the Parties, or COPs. There have been 28 COPs so far, with the latest in Dubai in December 2023. And while marginal improvements have been achieved, the world is still careening toward overuse and catastrophic warming, in part because voluntary action still dominates the agenda and regulations have been politically unpopular in a world focused solely on economic growth.

Things are getting really bad, though. The Institute for European Environmental Policy has calculated that from the beginning of recorded emissions in the 1700s up to 1990, the world emitted 784 billion tonnes of CO_2 into the atmosphere. In the 30 years since 1990, the world produced an additional 831 billion tonnes. To be clear, this means that we have emitted more carbon in the last 30 years than we did in the previous 290. Demand for energy and material resources from the planet is growing exponentially, and despite all the COP meetings and policies, we're way off track to decarbonize. Humankind is consuming far, far too much.

I would be remiss to ignore the fact that not everybody is equally responsible for the climate crisis. In a ground-breaking study published in *Climatic Change*, Richard Heede produced the most complete picture to date of which institutions have extracted the fossil fuels that have been the root cause of global warming since the 1850s. Rather than focusing on emissions by country, as many studies have done previously (China, the US, India and Russia top that list), Heede aggregated historical emissions according to carbon-*producing* corporations. The Carbon Majors Report concludes that roughly two-thirds of the CO_2 emitted since the Industrial Revolution can be traced to the 90 largest fossil-fuel and cement producers, most of which still operate today.

It took me a minute to pick my jaw up off the floor after reading that. So, not everybody is equally responsible. Some are way more responsible than others – and many of those corporations invest in sports to improve their public image (more on this later).

Those of us living in high-emitting countries per capita – I include myself in this, as a Canadian – are part of the problem. Our consumerist behaviors are too much for the planet to handle. Unfortunately, individual actions won't solve climate change.

In Canada (I'll pick on my own country here), the average person produces around 18 tonnes of CO_2 per year. The global average, for reference, is 4.8 tonnes. But most of the things we can do to reduce our emissions as individuals, like not driving cars, insulating our homes better, switching to heat pumps instead of furnaces, flying less (or not at all), are actions that we have little choice over because for most people, many of the low-emission options are too expensive or unavailable.

Living in a city center, where public transit is most developed, costs more than living in the suburbs. So people live outside the city and commute, often by car. This is by design; countries assess progress using gross domestic product (GDP), so from an economic standpoint it makes more sense to have people live in bigger homes where they buy more stuff, rely on cars, and live more carbon-intensive lifestyles. With GDP growth as the goal, it doesn't make sense to incentivize people to live in smaller residences, use public transit, and consume less in general. Government decisions around how our cities were built, where our energy comes from, and how our lives are structured, alongside very smart and successful advertising, have sold the "good life" as one of near-constant consumption.

But let's say you did everything right and lived the most sustainable life. Put solar panels on the roof, got a heat pump, installed better insulation, stopped flying and driving, reused and repurposed everything, used zero single-use plastics, stopped buying new stuff... it still wouldn't be enough. Individual carbon footprints, even big ones, like Canadians' 18 tonnes per year, account for only 0.00000000004 per cent of total global emissions. That's ten zeros. It's smaller than a rounding error. So, yes, we should do those things, and every bit counts. But even if you do, those individual actions won't fix it on their own.

The bigger battle is to learn about climate change, talk about it with the people you know, and demand better policies from our governments so that all these individual changes are easy, and so that everybody makes them. *That* would move the needle. For now, let's start on step one: learning more.

Climate change versus sports

Climate change could change everything about sports. In a similar way to COVID, this emergency is already impacting sports both directly and indirectly. As conditions worsen, and they likely will unless emissions are curbed considerably in the next 20 or so years, the sports sector will have to either retool and reposition itself, or risk falling apart completely.

The coming transition will be an uphill battle and comes with a steep learning curve. For decades, as professional and elite sports systems have established their business models and operations, they've taken the natural environment for granted. Big assumptions have been made about weather: summers are expected to be hot, but not too hot, and winters should be cold, with snow. Those assumptions aren't holding anymore.

But equally, the sports sector has made assumptions about acceptable land and resource use, most of which echoes larger business practices outside of sport. These assumptions, too, are now being challenged. Until a few years ago, nobody blinked at a professional team flying to nearby games, or questioned the lack of vegan food in sports stadiums, or gave a single moment of thought to the waste produced in sport. When I was a kid, people skied without worrying too much about the length of the season or the availability of snow – we just assumed it would be fine. In decades past, FIFA World Cups were held with minimal news coverage of their environmental impacts; for the 2022 FIFA World Cup in Qatar, overstated carbon neutrality claims were a leading news story in the weeks before the tournament. So what's changed?

For one thing, awareness of and concern about climate change have grown exponentially in recent years. In 2022, Meta and the Yale Program on Climate Change Communication ran a survey of more than 108,946 adults in 192 countries and found that most people in nearly every area surveyed (108 out of 110) said they are "very" or "somewhat" worried about climate change. Latin American and Caribbean nations had the highest percentage of concerned citizens, with Mexico, Chile, Costa Rica, Ecuador, Panama, Peru, and Colombia all having more than 90 per cent of worried respondents. Experience with recent disasters is driving concern in many regions, and mass-media coverage of the climate crisis is helping to drive awareness.

I studied and wrote my PhD at the University of Minnesota on the impacts of climate change on sports. It was a weird and unorthodox experience, in that my coursework was completed in four different schools across campus, and my dissertation committee had only two sports people on it; the rest were from the UMN Institute on the Environment, the Department of Forestry, and the Humphrey School of Public Affairs. A bit of an unusual mélange that confirmed what I suspected all along: the ideas we need to solve climate change won't come from siloed disciplines; they'll come from interdisciplinary engagement. Ann-Cecile Turner, long-time sustainability champion in the sport of sailing and campaign director at Sails of Change, once suggested to me that our role is to "build bridges in a world of silos". It felt apt at the time; it still does.

My early research goal was simple: I just wanted to make the case that climate change is already happening, and it's impacting sports in ways the sector isn't yet addressing. It wasn't a particularly challenging task because there's an overwhelming body of evidence that climate change is real (thank you, natural and physical scientists of the world!) and abundant examples of climate change impacting sports (as you'll see in this book). What was missing was a translator to take that information and interpret it for sports managers, coaches, athletes, and fans, helping them see the risks and how they can respond.

My interests and early lessons in this space led me to start theorizing around the relationship between sport and nature. I figured if few sports managers and coaches are thinking about climate change, maybe it's because they take the natural environment for granted. It was a relatively safe hypothesis: I was studying in North America, where most of the population can be described as out of touch with nature because our lifestyles simply don't require us to engage with it directly on a regular basis. Aside from farmers, foresters, fishers, landscape architects and engineers, not many people are monitoring nature and engaging with it on a daily basis. But outdoor sports invite people outdoors, and for many North Americans, time spent playing outside is their closest connection to the natural world.

My first big academic contribution was the "climate vulnerability of sport organizations framework", which took well-established concepts from climate literature and applied them to sport for the first time. The framework is simple: on one axis, you have climate impacts on sport. This is a measure of the likelihood and severity of possible climate-related risks. Potential climate impacts can be high – such as for an outdoor field sport in northern California, where wildfires are frequent and heatwaves are increasingly common, or low – such as for an indoor basketball gym in a city like Duluth, Minnesota, where flooding, extreme heat, seasonal changes, fires, and other major hazards are not (yet) common occurrences. On the other axis, you have adaptive capacity: the organization's readiness to adapt and resume normal activities with minimal loss or damage.

The most important part of the framework, in my view, is adaptation. There is so much sports organizations can do to insulate themselves from the worst impacts of climate change. In some ways, these adaptations resemble potential adaptations outside of sport. Much can be done to ensure athletes' health and wellbeing, deliver high-quality sports performance, protect business interests, and preserve the long-term viability of sports. But first, the risks have to be understood.

In the years that followed that initial framework, I spent time studying sports organizations' recovery processes after facing major climate hazards to learn what they might have done differently if they'd had more warning, more resources, or both. Many of the stories in this book were uncovered through that research.

This work has opened the doors to many lines of inquiry. Among them, I've begun researching structural inequities in sport and how these are exacerbated by the crisis, with my colleague Dr. Jessica Murfree. This work builds on the premise that the climate crisis is a threat multiplier – that is, it will exacerbate the existing threats and worsen inequities in every system. In the same way COVID showed us how fragile the healthcare system is, and how poorly certain communities are treated relative to others by governments and social services, climate change is laying bare all the cracks in our systems. This includes sports systems. The athletes, coaches, teams, and sports organizations operating at a disadvantage – whether financial, structural, social, or all three – are consistently worse off when faced with climate hazards than their more advantaged peers. More on this later.

I've also studied climate activism by athletes, a growing trend. Through interview-based research with outspoken professional athletes across a range of sports, I've started to pick apart what motivates some athletes to speak up on climate change, while others stay quiet. The results mirror some of the research on other forms of activism, but in many ways, climate activism is unique. I share more in Chapter 16.

Most recently, my work has veered toward understanding the roles of sustainability managers in sport: what do they know, what tools do they use, what solutions would make their jobs easier? A few years ago, very few sports organizations had a sustainability position on staff. When I would contact organizations, asking to interview someone (or multiple people) on staff about recent environmental challenges, I'd get puzzled looks, and would then be passed over to the facility management team. Over time, things changed. Interns were hired to perform sustainability tasks within operations departments, then those roles turned into full-time entry-level jobs. As the work got more sophisticated, the roles grew: from volunteers and interns, to managers, to chief executives. Now many major sport leagues have a full-time person or team working on sustainability: measuring emissions and waste, contributing to procurement efforts to find more environmentally friendly suppliers, supporting the communications and marketing departments to talk about these efforts publicly. The shift happened fast. It's still going on. What a great time to be a sports fan.

————————

My gap year as a ski bum became the start of a career in academia and advocacy that I never could have predicted. I've since founded the Sport Ecology Group, an international consortium of academics aimed at studying the impacts of climate change on sport, the many ways the sports world can reduce its environmental footprint, and how sport can increase its influence on environmental policy and public opinion.

I've spent time helping sports organizations large and small navigate their climate vulnerability and adaptation options. I've spoken to journalists at media outlets ranging from ESPN to *Time* Magazine to the BBC about climate issues in sport. In doing this work, I've joined a large and growing group of people who are deeply concerned about climate change and who use sports as a platform to effect change.

Dr. Michael Mann, one of the world's pre-eminent climate scientists, used a hockey stick to explain the way the global mean temperature is increasing. The hockey stick-shaped curve appeared in research on the science of ozone depletion, and Jerry Mahlman, Director of Princeton University's Geophysical Fluid Dynamics Laboratory, gave it that name. When I asked Mann why he and his team used the hockey-stick analogy to explain warming trends, he replied, "I think that analogies are very helpful in explaining scientific concepts to a lay audience. And sports metaphors are always especially accessible and sticky. So I think the name helped the conceptual advance reflected by the hockey stick penetrate into our mainstream public discourse." His hockey stick turned heads and commanded national attention in the media. But he wasn't the first to figure out that sports stories and analogies are an effective tool for climate communication.

Jeremy Jones is probably the best example of a sports climate communicator. A world-famous snowboarder and one of the most accomplished action sports athletes of all time, Jones founded the non-profit Protect our Winters in 2007. POW, as it's affectionately called by the winter-sports community, is the world's only winter athlete-advocacy platform aimed at lobbying for environmental protections. With chapters in 15 countries, plus a regional chapter for Europe, POW regularly sends professional winter athletes to speak in schools and government hearings to raise awareness about climate change and its impacts on the ski industry. I've personally sat in meetings with Canadian government officials and POW representatives, and watched them at work: they lean on personal stories of climate-related losses matched with the latest science, fact-checked by members of their Science Alliance. It's impressive to see them do what they do.

The examples of sports and climate advocacy stretch beyond the winter sports scene. In Atlanta, former NFL (National Football League) player Ovie Mughelli began advocating for climate action when his twin daughters were born early and held in the NICU for two weeks – not because they were unhealthy, but because Atlanta's air quality was so poor that the pediatricians determined it was unsafe for the babies to be outdoors. Mughelli has become one of the most vocal athletes in America advocating for climate action. In the past few years he's launched the Ovie Mughelli Foundation to raise funds for climate education, and created a comic book series called *Gridiron Green* about a running back turned environmental superhero who's entrusted with protecting Planet Earth.

In motorsports, Lewis Hamilton, seven-time World Champion of Formula One racing, recognized the damaging effects of climate change, and changed his diet and investment portfolio to participate in climate solutions. When asked by the BBC about his vegan diet in 2017, Hamilton said, "As the human race, what we are doing to the world ... the pollution [emissions of global-warming gases] coming from the amount of cows that are being produced is incredible ... The cruelty is horrible and I don't necessarily want to support that." Despite facing criticism over his expensive and high-emitting lifestyle flying around the world to compete in motor racing events, Hamilton maintains that he's actively working to raise awareness of climate change, and using his platform to do so. This includes being an early investor and team owner in the Extreme E racing series, which aims to draw attention to parts of the world that are heavily impacted by climate change and accelerate the technological shift toward green transport.

The sports sector has the ideal language and storytelling capacity to be engaged in the climate fight. This is an industry where the status quo is a relentless pursuit of excellence. Fans and athletes alike expect to smash records each year, pushing the boundaries of the human body to its upper limits, and then do it all over again the next season. Chasing improvement is good, but not good enough − sports people are looking for results. This is also an industry where people know it's impossible to win every time, and losing is part of the game. There will be losses in climate change. Big ones. There already are. That doesn't mean we stop and give up. Losses serve as fuel and motivation for future wins, and the same mentality can be applied to climate action − that is, if the global sports community is ready to have that conversation.

I'm not writing this book to scare or shame people into "doing something" about climate change, although it would be cool if reading this inspired you to act. I'm writing to share stories of climate change in the sports world, to explain how this crisis is already impacting the places we play and to show that adaptation and mitigation are possible.

I also want to make it clear that this book is not meant to draw attention away from the ongoing life-and-death crises occurring in the developing world, where communities are suffering the worst impacts of the climate crisis, particularly in the Arctic, along the equator, and in historically exploited nations where resource scarcity is severe.

Often, when I speak publicly about climate change and sports, I get critiqued on social media by people asking why airtime is being devoted to sports climate stories instead of more urgent and dire situations elsewhere in the world. While they're right that the Western media's climate coverage is inadequate, I don't believe the goal should be to focus only on some climate stories and not others. The goal must be to talk about the climate crisis more across all channels, and on all topics, period.

At the beginning of 2023, I was encouraged to see Rebecca Solnit open an inspired essay in the *Guardian* with:

> Every crisis is in part a storytelling crisis. This is as true of climate chaos as anything else. We are hemmed in by stories that prevent us from seeing, or believing in, or acting on the possibilities for change. Some are habits of mind, some are industry propaganda. Sometimes, the situation has changed but the stories haven't, and people follow the old versions, like outdated maps, into dead ends.

The fact that people follow and care about sports makes for an easy way in to conversations about harder topics. This tactic has been adopted before to address gender inequities and racial injustice. Think of Megan Rapinoe and the members of the FIFA World Cup-winning US women's national team who publicly pursued equal pay with the less successful men's side. Or recall Billie Jean King in the previous generation. In the summer of 2016, Colin Kaepernick and a slew of other athletes, including the whole Minnesota Lynx roster of the WNBA (Women's National Basketball Association), took a knee and wore Black Lives Matter shirts on the sidelines to protest against ongoing racial injustice, especially police brutality. The movement later grew in 2020 following the George Floyd murder and resulted in league-wide cancellations of games in the NBA (National Basketball Association), MLB (Major League Baseball), and MLS (Major League Soccer).

What's different with the climate crisis, and what will hopefully mirror the COVID response, is that the institutions of sport themselves responded to COVID, not just the athletes who stuck their necks out to advocate for a worthwhile cause. Sport provides an avenue and a platform for these tougher conversations. With an estimated 3 billion fans worldwide, which works out to roughly a third of the global population, the sports sector could be a direct and practical mouthpiece for communicating climate issues to large and diverse audiences.

In this book, I recast the last couple decades of major climate stories through a sports lens and share stories about how ski resorts handled short winter seasons, how a championship marathon was held in the middle of the night to avoid the heat, how fields have been repeatedly flooded, how sporting calendars have been disrupted, how golf courses have eroded and begun falling into the sea, and how athletes have suffered avoidable heat-related illness. None of these issues is more or less urgent or important than the other, and the order of the chapters is inconsequential.

Some of the stories happened as a direct result of the climate crisis. It's possible to differentiate those climate hazards that occur due to climate change from those that would have happened naturally, without climate change, thanks to attribution science. In attribution science, researchers calculate the statistical likelihood that a given climate hazard such as a wildfire, a heatwave, or a hurricane, would have occurred in a hypothetical version of the world that was not warmed by human activity. If the likelihood is extremely low, we can reasonably deduce that the event was caused, at least in part, by climate change. When these "unlikely" climate hazards continue to happen, we can be even surer that the climate crisis is part of the problem.

In this book, some climate hazards have been directly attributed to climate change. In other cases, the climate hazard was not attributed to climate change, but the science has made it very clear that future warming will make those hazards more common, more severe, or both. So we ignore them at our peril. What these stories have in common is that they are cautionary tales of the future in a climate-changed world if warming goes unchecked.

The first half of the book introduces athletes, teams, and events that have been directly impacted by climate hazards. These stories come from all over the world, and no sport is immune. Together, they illustrate that climate change is happening right now. It's been happening for a while. This is not some future problem, it's urgent and immediate. I also paint a picture of what sport's impact on the planet currently looks like, so that later in the book, I can focus on the sports sector's work to fix it.

Most of this book is grim. I'm not big on the doom-and-gloom narratives, but it's important to tell these stories honestly and show that the climate crisis is an ongoing disaster. So I didn't shy away from the dark side. But I did work hard to find the light and the opportunity in these stories because I know how easy it is to shut down when all the news is disheartening.

The last five chapters are my attempt to craft a narrative of what our future could look like, if the cautionary tales are heeded, appropriate changes are made, and the sports sector wields its exceptional influential power to inspire other sectors to follow along.

Most people are surprised to learn I'm not much of a sports fan (though my dad and husband may cringe when they read this, because they really did try – sorry, guys). I became interested in sports in the same way an atheist becomes interested in religion: because it's powerful, because it brings people together, because there's so much potential to do good in this space.

Nelson Mandela famously suggested that sport has the power to change the world. I mostly agree with him and we've seen it work before, but I'd like to add a caveat. Sport has the power to change the world *if* the sector intentionally organizes itself to do so. Changing the world doesn't just happen. For the world to slow, stop, and reverse the climate crisis and preserve the lives, livelihoods, and games we love, it's going to take all (or most) of us working together. Climate action is a team sport.

CHAPTER TWO

HEAT CHECK

Jordan McNair was not a regular kid. He stood out for his size at six foot five and 325 pounds, with a size 16 shoe. He also stood out for his talent as a football player. Jordan had spent his high-school playing career being watched and heavily recruited by Division I football programs across the US. Coaches who worked with Jordan were quick to comment that he was a quiet kid and very coachable, with huge potential.

On May 29, 2018, Marty McNair got the call every parent dreads. There had been an emergency. Jordan had suffered a seizure during an off-season practice. Recalling that day, Marty explained, "At first, I didn't think much of it, Jordan was a healthy guy, hadn't been in the hospital since the day he was born. He was in good shape and so this just didn't activate panic for me. I had no idea what was going on."

When Marty and his wife Tonya arrived at the hospital, Jordan was in a Bair Hugger™, a medical device which wraps around the body, as close to the skin as possible, to regulate its temperature – kind of like a heating blanket, but it can be used to warm or cool the body. At the time, neither parent knew what a heat-related injury was but, seeing him in the hospital bed, the severity of the situation became apparent.

In the days that followed, as Jordan's condition deteriorated in hospital, and as he underwent an emergency liver transplant, Marty started asking questions. "I wanted to know what the emergency procedures at the university were supposed to be, what the coaches knew about heat injuries, what the trainers knew, and whether this had ever happened before."

At the time, Marty and Tonya didn't know there had already been 30 deaths among college football players since 2000 due to heat-related illnesses. They didn't know Jordan was becoming a statistic. "It felt like a freak incident. And to some extent it was, but as I kept asking questions, I learned that this type of incident could've been prevented."

Jordan McNair died in hospital 15 days later, on June 13, 2018. The cause of death was heatstroke. In the following months, Marty's questions turned into full-blown investigations and a justifiably hefty lawsuit.

Heat illness is entirely preventable. To have an athlete die of heatstroke is not something that should happen if emergency response protocols are in place. An external review of the McNair incident conducted by Walters Inc. – a private consultant with expertise in sports medicine – found that in Jordan's case, an emergency response protocol was not followed and the care the university provided was not consistent with best practices. The steps to prevent heat illness were not taken, it was not identified when Jordan started showing symptoms, and the protocols to address heat illness were not followed: athletic training staff did not take his temperature and implement a cold-water immersion treatment fast enough. In addition, the review laid out the timeline of the incident and a toxic team culture as being relevant to the outcome.

The day went something like this. On May 29, the first athletes walked onto the field at 4:09pm and the workout began a few minutes later. At 4:40pm, the players began a set of ten 110-yard (100m) sprints, a conditioning test. By the seventh sprint, around 4:53pm, Jordan showed visible signs of fatigue, slowing down considerably and missing the allotted 19-second pace per sprint. Two fellow players who were interviewed for the investigation reported athletic trainer Wes Robinson yelling across the field to "get [Jordan] the fuck up", and "drag his ass across the field".

After the sprints, Jordan was seen by two assistant athletic trainers on the field who documented his symptoms as including back pain and cramping, hyperventilation, profuse sweating, and dizziness. It took 34 minutes before Jordan was removed from the field and taken to the athletic training room where he was hydrated and given cooling towels. Thirty minutes later, a full hour after his initial symptoms showed during sprints, Jordan experienced what trainers called a drastic "mental status change" and began yelling. The head football athletic trainer noticed the change and instructed his colleague to call an ambulance.

The ambulance wasn't called. Instead, a call was made to the team physician, who reiterated the advice to call an ambulance. At 17:55, emergency services were called. Moments later, Jordan had a seizure on the training table and struggled to breathe due to an airway obstruction (a mucus described in the report as a brown foamy sputum). By the time they arrived, found Jordan's location (there were issues with flagging down the paramedics), loaded Jordan's gurney, and drove to the hospital, nearly two hours had elapsed since the first symptoms presented.

Four players were interviewed for the investigation. Their comments, which were paraphrased in the Walters report, show that players had little trust in the training staff, and that they were encouraged to have "blind trust" in the coaching staff. One player expressed concern that all the football staff were present on that day but still managed to miss the signs of a player in distress. The player was suggesting that if the emergency response is slow when all the coaches, all the trainers, and all the strength and conditioning staff are on site, imagine what might happen if a player showed signs of heat illness on a day with fewer staff present. One player said that coaches preached a "no quit" mentality and that it was viewed as weak to seek medical assistance for any reason.

Following the report and internal investigations, Wallace D. Loh, the university's president, fired Coach Durkin. In a letter to the university community, Loh wrote that "a departure is in the best interest of the University". Soon after, the two athletic trainers who had attended to Jordan were also fired. The strength and conditioning coach who supervised the practice where Jordan collapsed resigned.

Marty McNair was relieved at the news of staffing changes, but it took another two years to reach a settlement with the university, despite Loh acknowledging the university's moral and legal responsibility for his son's death.

When I spoke to Marty in 2021, his loss still felt fresh. Together with his wife, Marty launched the Jordan McNair Foundation to educate parents, coaches, and athletics staff about heat-related illnesses and to distribute resources for treating it, such as cooling baths and hydration units. At the time of our call, the Foundation had donated more than 200 cooling tubs to high school and college programs across a range of sports, from Florida to Alaska. The Foundation also started working with legislators to get bills passed: one at the City of Baltimore, and the Jordan McNair Act in the State of Maryland, which both provide support in case a student athlete feels uncomfortable: they can contact a hotline anonymously to report discomfort or fear, without worrying about retribution. Separate federal bills were being drafted by Senators Richard Blumenthal of Connecticut and Cory Booker of New Jersey on student athletes' bill of rights, which Marty said would include health and safety considerations, and protections against retribution if athletes reported bad staff. However, a student bill of rights will only be effective if athletes are aware it exists and feel comfortable using it.

Beyond these laws, which are a good step in the right direction, Marty was very clear: not enough is happening on this agenda. Too few parents know about the risks of heat-related illness. "When Jordan was growing up, nobody ever told us about heatstroke, or any injury at all. And we never thought to ask the questions. And now that I think about it, never did we have a safety conversation at all, on any topic."

Marty also highlighted that the National Collegiate Athletic Association (NCAA) does not value athletes' lives very highly, and suggested that protecting athletes from fatal outcomes like heatstroke is not a top priority. As of 2019, the NCAA values its student athletes' lives at about 10,000 US dollars (£7,880). "That's what the life insurance policy was for Jordan. Ten thousand. That's what they think your child's life is worth."

Weeks after I spoke to Marty, Drake Geiger, a 16-year-old high school football player in Omaha, Nebraska, collapsed on the field during a practice and later died in hospital of heatstroke. On the same day, August 10, 2021, Dimitri McKee, an 18-year old from Montgomery, Alabama, passed out during practice and was airlifted to hospital where he later died, also from heatstroke. Meanwhile, in Georgia, two high school basketball coaches were charged with murder and child cruelty relating to the August 2019 death of Imani Bell, by heatstroke, at a pre-season outdoor basketball practice in Jonesboro.

The need for a speedy response

Exertional heatstroke (EHS) is one of the most common causes of death in athletes. It also presents unique challenges due to the accelerated timeline of heat illness and the potential for conditions to worsen quickly. Dr. Kristen Kucera, Director at the National Center for Catastrophic Sports Injury Research, explained to me that if an athlete is experiencing exertional heatstroke, and cooling interventions are not immediately applied, the likelihood of organ damage, morbidity, and mortality increases significantly after just 30 minutes, which is faster than most EMS transports can be called, arrive on scene, transport the athlete to hospital, and have them seen by the emergency department physicians.

So how do bodies reach this point? In a healthy natural state, the body stores a lot of heat. And in physical activity, the body generates more heat. As large skeletal muscle groups contract (think of your legs, your back, your core), they use quite a lot of energy, some of which goes toward doing muscular work – or achieving the movement you are doing – while the majority of that energy is released as a by-product, heat. To avoid overheating, the body needs to shed that extra heat.

The body has two ways of shedding excess heat: sweating, and evaporating sweat off the skin. If exercise is done in a warm environment, one of our heat transfer pathways – from the skin to the environment – is taken away because the difference in temperature between the skin and the air is not significant enough. When air temperature reaches the high 30s Celsius (100 Fahrenheit), the temperature gradient between the skin and the air is lost, because most bodies have average core temperatures around 37 degrees Celsius (99 degrees Fahrenheit). At that point, the only way you can lose heat is through evaporating sweat. If you throw high humidity into the mix, the evaporation of sweat is compromised, which is when you really start running into trouble. So when you see athletes competing in warm, humid environments, like Tokyo during the 2021 Olympics, or the Southern US in the summertime, that's when problems arise.

Exertional heat illness implies that the primary source of heat is physical activity, not the ambient heat, although that contributes to the problem. There are levels of exertional heat illness. The first and least severe type of exertional heat illness is heat edema or swelling, typically of the hands and feet. Heat can also give rise to heat exhaustion, which is where you start having signs and symptoms of heat illness. Signs and symptoms of heat exhaustion include fatigue, profuse sweating, ataxia – a lack of coordination – and nausea. Usually, these symptoms are all driven by rising body temperatures, and aggravated by hydration status. At this stage, a break from play, some water or sports drink, and a few minutes out of the sun, is enough to cool the body and curb the symptoms.

If the body doesn't get a break from the heat and continues warming, heavy sweating can lead to sodium loss. If this happens, individuals may also experience heat rashes, cramps or muscle spasms, and fainting. At the more severe stages, excess heat in the body can result in heatstroke. All these types of illnesses, taken together, are classified as exertional heat illnesses.

How the body can adapt

Let's focus on exertional heatstroke, the most severe and potentially lethal form of heat illness. In exertional heatstroke, the body redirects blood away from the center of the body toward the skin to aid heat dissipation, helping the body to sweat better. That process can help with cooling, but it starves your gut of blood, and therefore oxygen delivery. And so, the heart will adapt by pumping harder, trying to get more blood to where it needs to go. The endothelial barrier of your guts, which are the cells that keep all the nasty stuff inside, are compromised by the combination of low oxygen and high local tissue temperatures.

When the endothelial barrier is starved of oxygen and starts overheating, it gets looser, increasing gut permeability, so the toxins that help your body break down food can leak out, enter the circulation system, and set off a cascade of potentially catastrophic events. This is called the leaky gut hypothesis. In Jordan's case, it took less than an hour for this process to begin, and several days for it to escalate to the point where a liver transplant was necessary, which ultimately proved insufficient to save his life. Too much damage had been done.

Now, athletes can have some resilience toward exertional heatstroke if they're well conditioned and adapted to the environment. Athletes become adapted to hot environments, particularly while exerting themselves at high levels of physical activity, either through a process called acclimatization, or through acclimation. These processes are similar but with one important distinction: acclimatization is the process of adapting to a naturally occurring environment, while acclimation is when you purposely expose yourself to doses of heat exposure in an artificial environment such as a climate chamber. Acclimation and acclimatization can also work the other way, getting the body used to cold temperatures, or preparing for altitude.

If athletes have good resources and access to climate chambers, sports scientists can help them achieve a good level of adaptation within the space of a week. This is accomplished by exercising at a high percentage of their maximum capacity, perhaps in a really hot chamber, or perhaps in a humid environment, created artificially in a climate chamber. According to Professor Ollie Jay at the University of Sydney, spending 90 to 120 minutes in a climate chamber for seven consecutive days is enough to see the body respond and adapt. Those adaptations include a dip in resting core temperature by a few tenths of a degree, and a slower heart rate for the given activity in that specific environment.

While the athlete exercises inside the climate chamber, their body begins to produce more plasma, increasing the total volume of blood, not only so that their heart doesn't have to work as hard to pump blood to the parts of the body that need it to perform the exercise, but also so that there's extra blood available to be dispatched to the skin when the body heats up. It also becomes easier to sweat, so the maximum sweat level and the amount of surface level that you can saturate with sweat goes up. Those are the main adaptations achieved after only seven days. To guarantee full acclimation, says Jay, "do what I said for 14 days instead of seven, but seven is still pretty good." Even three or four will give you some benefits.

With most athletes and teams, access to a climate chamber is out of reach, so they turn to acclimatization. This is when teams travel to a training camp in a hot place that has a similar prevailing climate to where they're going to compete for about a week or two beforehand. These environments provide the same benefits as acclimation. The athlete should receive progressive exposure to the environmental conditions in a natural setting, not a heat chamber. The key words here being "should" and "progressive".

Through either acclimatization or acclimation, athletes are able to sweat more, tolerate higher temperatures, achieve greater cardiovascular stability, and maintain a slightly lower core body temperature. The other thing that athletes benefit from, if they're adapted to the environment, is something called chaperone cells. These are cells that help increase resistance to heat stress symptoms, reduce the risk of cell damage with high temperatures, and contribute heat shock proteins to improve the body's capacity to cope with heat. Chaperone cells have a wonderful protective effect. So, in athletes that are in competition shape and adapted to the environment, you see less damage to the cells at high temperatures than you do in people without these cells. There's evidence to show that you have an expansion in those chaperone cells with repeated and progressive increases in heat exposure. Again, the key word is "progressive".

When athletes have been away from hot temperatures for a while, the benefits of any acclimatization they acquired in a previous season will fade. According to Dr. Rebecca Stearns at the University of Connecticut's Korey Stringer Institute, this regression can happen in as little as two weeks. Regular exposure to heat – meaning regular exercise in hot conditions, typically twice a week or more – is necessary to keep the acclimatization benefits.

It comes as no surprise, then, that the first week back to practice after a prolonged break is the most dangerous in terms of heat risks for athletes. For American football in particular, the pre-season lands in midsummer in the hottest conditions, only making things worse. Not only are the athletes' bodies not ready for any kind of hard workouts, they also aren't acclimated to the heat. Progressive exposure would dictate that some degree of physical fitness be first developed indoors or in conditions with more moderate temperatures, then moved outdoors for short periods of time and gradually scaled to longer outdoor practices in the heat.

But in a changing climate, people in all corners of the world are facing unprecedented temperature extremes for which nobody is acclimatized. This is where mid-season blips can occur. Temperatures that are much hotter or more humid than a location's average are considered "extreme", and we're expecting more extreme temperature days on every continent. When you get more than three consecutive days of extreme temperature back to back, they become known as extreme heat events or heatwaves. In some parts of the world, we're seeing so many more extreme days, they barely earn the moniker "heatwave" anymore. Summers are just hotter. That's the new normal.

It's important to highlight here that exertional heat illness is not the same as heat illness that is experienced by average people who are just going about their everyday lives or suffering during heatwaves. Typically people who end up getting sick and fatigued in hot conditions, or who actually perish in these circumstances, do so because they have a low adaptive capacity to deal with the environment. In heatstroke cases that happen outside sport – or other physically active environments like the military or firefighting – the person gains heat from the environment because the external temperature is hotter than their body temperature. This can lead to a catastrophic cardiovascular event that occurs because the heart is working hard to send blood to the skin, to the overheating organs, and to the brain. It's got to pump more to achieve the same circulation, and it can become overworked and shut down.

One question Professor Jay said he often gets is whether regular people will automatically acclimatize to heat if they live in hot climates. His answer is straightforward:

> I doubt it. When we acclimate athletes in a chamber, we're exposing them to like 48 degrees Celsius [118 degrees Fahrenheit] and 50 per cent humidity, and I'm getting them to exercise at 70 per cent of their maximum capacity for an hour and a half, seven days straight, or 10 days straight. That's a huge hammer that we use to get them ready, and to activate the full suite of physiological adaptations that allows them to perform in hot conditions.

Air conditioning contributes to this, too. In places where many people don't have air conditioning, like Europe, Africa, and South America, they may get some benefits of acclimation through their everyday lives. Not the same benefits as athletes would achieve through an acclimation protocol, but some. In places like North

America and Australia where air conditioning is pervasive, those acclimation benefits are not achieved in the course of everyday life because even if it is hot, people don't expose themselves to the heat. Jay uses the analogy of a gym to explain this: "If my office is next door to a gym, I don't get big muscles by just sitting in my office next door to the gym, I've got to go in and expose myself to the stimulus in order to get that result." To benefit from heat acclimation, you need to intentionally acclimatize.

Most of the research on acclimation and acclimatization has been done on adults, so the jury is still out on kids, "but it's unlikely that there's a difference," says Jay. His lab is in the midst of a large externally funded project to develop an evidence-based heat policy for child and youth sport in Australia, but collecting the data to back that up is a tall order. "There is really no data for kids yet. Certainly no good data for kids. So we're generating it." However, while we may not have the child-specific data to make distinctions between children and adults, it's still clear that generally kids should not be playing outside in extreme heat.

Indoor sports

In the absence of good air-conditioning systems, athletes of indoor sports can also be affected. In 2016, Australian professional netball player Amy Steel collapsed in the parking lot of Shepparton arena in Victoria after playing a full game in brutal, unconditioned heat. Despite being a low-stakes, pre-season exhibition game, it was her last.

The temperature outside was 39 degrees Celsius (102 degrees Fahrenheit) that day, and youth sport in the area was canceled due to the heat. But because so many tickets had been sold to the netball game, officials decided this game would go ahead. In the hours leading up to the game, the athletes were taken to the center of town to sign autographs and make appearances alongside AFL (Australian Football League) players who were also in Shepparton for exhibition games; these small-town game opportunities are important opportunities to promote the game in regions where professional sport doesn't exist.

When Amy and her team got to the arena that night, it felt like walking into a wall of heat. She recalls asking the coach whether someone had checked the temperature in there, and was reassured it was fine. This was before any fans had come in, so the space was still mostly empty. Nonetheless, because of the obvious heat, coaches made plans to rotate the athletes regularly throughout the game.

Amy plays goal defense, the position with the highest intensity repeat effort. Several studies have used heart-rate monitors and satellite mapping of a netball court to demonstrate that athletes in goal defense burn the most calories because of the repeat bursts of movement. By halftime, she was sweating profusely and feeling exhausted. She was ready for a break. But it didn't come. Every other player was taken off the court, except her. She looked into the crowd and saw that her parents, who had traveled to watch her play, had left early. She would find out later that they had left because the heat had been unbearable in the stands.

"At this point, I was thinking, this is a bit of a compliment. It's a good thing, the coach trusts me. I thought I'd just toughen up and push through. I wasn't feeling well, but there was no way I was going to say to my coach, oh, can I please come off?"

She kept playing. When the game ended, the team spent a few minutes stretching, then went back into the arena to sign autographs. The athletes hadn't yet rehydrated or eaten, nor had they showered. The fans were also visibly uncomfortable in the heat, Amy recalls. But this was the job. So she smiled and signed autographs for 20 minutes.

By the time she got into the cooling bath, it became clear that something was really wrong. It wasn't just that she was sweaty and tired, but the water didn't feel cold.

I started to feel weird temperatures. I got in the icebox and I distinctly remember asking my friends, the other girls who had hopped in the ice bath, I was like, "Is this cold? Does this feel cold to you?" And they said, "Yeah, of course it's cold. It's an ice bath." That struck me as weird, so I got out and went to the shower. At this point, I was feeling a bit loopy, I wasn't really sure what was up and down, and I couldn't work out how to have a shower.

She walked out of the shower, got dressed slowly, and made her way out to the parking lot with her teammates. That's when she collapsed.

The team trainer took her to the hospital, but it was full, it being a small town with limited medical personnel on an absurdly hot day. She wasn't going to get seen that night. So they took her back to the hotel, administered some first aid, got her fluid levels up, and sent her to bed.

Back in Adelaide the next day, she was feeling OK, so she went in to work.* And there, she collapsed again. Back to the hospital she went. This time, they ran every test under the sun. But nothing was coming up with definitive answers, except for a high creatine kinase (CK) marker, a classic sign of muscular degeneration or trauma. So the tests kept going.

Ultimately, it took more than nine months for Amy to recover from the heatstroke. She was in and out of hospital, bedridden. Her immune system weakened considerably. It took months for her to realize that her sports career was over. Now, more than six years later, rigorous physical activity is still out of the question, as is any exposure to heat.

---

*Women athletes are notoriously underpaid in professional sport, so Amy, like most of her team, had a full-time job on top of her full-time sport. Gender inequities are rampant in sport, but that's a rant for another book.

The factors at play

Pinpointing exactly who is at risk from exertional heat illnesses can be tricky, because there are so many factors at play. Temperatures, humidity, air velocity, an individual's acclimatization levels, body mass, sex, where a person is exercising (urban or rural locations), and more.

Epidemiology studies show rates of exertional heat illness in the US have doubled since 1975, though it's possible some cases have been misidentified as heart failure or other circulatory issues – so the number could be even higher. More than 70 per cent of observed cases have been men, though it's possible the football statistics are skewing the overall numbers, as American football is the most risky sport for heat illness due to the season, the equipment, the body types it attracts, and the types of workouts they conduct – high intensity, short bursts. Some research into gender differences demonstrates that women sweat less than men, on average, because size and hormones play a role in sweating. The bigger the person, the more heat the body will produce when exercising, and thus more sweat. So, if men are bigger than women on average, they'll sweat more. Women also have estrogen in their systems, which promotes a lower average body temperature, so a woman's body has to heat up more than a man's before she will start to sweat.

Location is another piece of this puzzle. In cities and urban areas, concrete and asphalt used for roads, and steel and cement used in buildings will absorb – rather than deflect – heat, and as a result, these areas are warmer. If you've ever tried to walk barefoot along a roadway in the heat of summer, you'll recognize this. The dark surface burns the feet, much more than grass or other natural surfaces. During the heat dome event in Oregon, Washington, and British Columbia in June 2021, I saw a few TikTok videos of people frying eggs on driveways or roads – the surfaces were that hot. These spaces also tend to have fewer trees, which can help to reduce temperatures. They are typically more densely populated too, so there are more people using energy and moving around.

All these factors combine to produce what's called the "heat island effect", which describes the markedly higher temperature in urban areas, sometimes up to 15 degrees Celsius (60 degrees Fahrenheit) above the temperatures on the same day and time in a nearby rural area. As a result, playing sport on a hot day in the city can be much more dangerous than playing in the countryside.

According to every expert I interviewed, exertional heatstroke is entirely preventable if the right adaptive capacities exist, and if heat stress is interrupted early with cooling, proper hydration, and activity breaks. Dr. Kristen Kucera pointed to three levels of prevention, typically applied in epidemiology: primary prevention, which is the prevention of heat illness in the first place; secondary prevention, which is the fast identification and intervention of cases when they do arise; and tertiary prevention, which is focused on managing cases to ensure minimal long-term issues or complications.

Primary prevention

For primary prevention, one tactic is to adopt heat policies. There are a few types of policies that can be developed to prevent heat illness in the most dangerous environmental circumstances. For instance, acclimatization policies dictate return-to-play standards for the first five days of practice in a new season or after a break from routine exercise. This return-to-play period is especially risky since athletes have lost their acclimatization benefits, so this type of policy can specify staged re-entry plans, what types of environments athletes will play in (for example, indoor practice the first week), the equipment that will be worn, and the content of the practice. Dr. Zachary Kerr at the University of North Carolina has studied the efficacy of heat policies and found that mandated guidelines for heat acclimatization in the first two weeks of pre-season high school football practices was associated with a 55 per cent reduction in cases of exertional heat illness.

Other policies focus on wet-bulb globe temperature (WBGT)*
readings. The WBGT is a measure that considers the effect of
temperature, humidity, wind, and infrared radiation on the
body, and can be found on weather apps or specialist websites.
Policies based on WBGT will typically provide a tiered approach,
listing playing adjustments for temperatures above a certain risk
threshold, and stipulating a no-go temperature for practices and
competitions when WBGT is especially high. These thresholds,
and the accompanying adaptations, have to be specifically crafted
for each sport and context: different levels of athletes have different
needs, different experience with managing their bodies in heat,
and different acclimatization levels.

Some states, such as Arizona, California, Mississippi, and New
York, have adopted statewide policies to govern heat at the
elementary and high school levels of sport. At the college level,
the NCAA (National Collegiate Athletic Association) has a heat
policy but it's more of a guideline and not always actioned. At the
professional levels in the US, policies exist but are not binding and
thus not always followed.

Professor Ollie Jay has developed several competition-specific
heat policies in Australia, where exertional heat illness is a top
concern. One of the most visible policies he created is for the
Australian Open tennis tournament. In 2023, the heat policy
was enacted in the first days of the tournament and all outdoor
play stopped. A decision to stop play in a high-stakes televised
tournament is not easily taken. Jay's policy had to consider not
only the athletes' wellbeing, most of whom land in Australia weeks
before the tournament to acclimatize at smaller events, but also the
fans, volunteers, and staff on site. It gets even more complicated

---

*Wet-bulb globe temperature combines humidity and temperature into
one value. This is typically measured by a thermometer covered by a water-
soaked cloth, and it provides an indication of how much evaporation is
possible in a given set of heat and humidity conditions.

when broadcast contracts are considered, and the expensive consequences of delaying live television programming are factored in. So this particular policy has higher "no-go" temperatures than policies developed for youth sport or a recreational setting.

Another primary prevention tactic is to provide breaks in the play, and build in cooling plans. Tennis is a good example of this, because there are many breaks built into the game. After every other game, there's a 90-second break; at the end of a set, a 120-second break. Jay and his team have been exploring what can be done to cool athletes down in those 90- and 120-second windows in two different environments: one hot and dry, similar to the conditions at the Australian Open, and one hot and humid, in a heat chamber at the University of Sydney. They found cooling towels with crushed ice inside placed over the neck and a damp cold towel over the legs provided sufficient cooling in those short windows of time. The Australian Open has since incorporated those strategies into the tournament.

Jay did similar research for the National Rugby League in advance of the World Cup during the Australian summer, and found an extension of the halftime break from 12 minutes to 20 made a big difference in cooling.

Misting fans and portable air-conditioning systems also work. At the Tokyo Olympics, the hottest Games on record, Spanish tennis player Paula Badosa had to be wheeled off the court partway through her quarter-final match after experiencing a heat injury. The next day, organizers installed portable air conditioners on the courts so athletes could cool their bodies down on the bench in between games. Misting fans and cooling tents were set up on the sidelines of several courts, and outdoor playing spaces were modified to ensure athletes were out of the sun and had adequate access to cooling. Obviously, these solutions are less than ideal from a sustainability standpoint, owing to the energy and water use, but safety comes first.

But sometimes, extra breaks and cooling options aren't enough, and rescheduling is required. As the Tokyo Olympics wore on, and temperatures stayed high, several events were shuffled around to avoid the midday sun. Dr. Paolo Emilio Adami, Medical Manager at World Athletics, spent the two weeks of the Games at the track, leading medical efforts. Despite only having races up till 11am, and then again from 3pm onward, the sun was punishing: "If you were in the shade, it was fine. Under the sun, it was just unbearable. We were ready for it, we expected the heat, but there were also some changes to the schedule that had to happen." Most famously, the women's gold-medal football game between Canada and Sweden was rescheduled from an 11am start time to 9pm to avoid the heat, and was moved from Tokyo to Yokohama because the revised start time would have clashed with the athletics program in Tokyo.

Finally, there are preventative measures that can be taken to improve the athlete's ability to sweat. One is acclimatization, discussed earlier. The other is wearing less equipment and heavy clothing.

Professor George Havenith is a world-leading researcher in environmental physiology and has spent 30 years studying heat and vapour transfer. His research has demonstrated how sweat passes through – or gets stopped by – various types of fabrics and materials used in sports gear. These materials can either facilitate or impede the body's capacity to cool itself down.

According to Havenith, wearing light equipment – or training without equipment in very hot conditions – is key. Foams and plastics don't let moisture through at all, so that is a challenge in American football, for instance. In most sports, Havenith explains, "the only uniform is fabric, and the type of fabric matters. In recent years, sportswear brands from Nike to Lululemon have touted the sweat-wicking capacity of their clothes, which is a process that draws moisture from one side of the fabric nearest your skin, and disperses it so it spreads across the fabric, making it evaporate more easily."

Polyester, a synthetic fabric derived from petroleum, wicks sweat better than other fabrics, and can therefore be more comfortable for athletes. In contrast, cotton absorbs more sweat, so it can be heavier and wetter to wear, but it's a more sustainable material. Each one has benefits, but Havenith was quick to say there are "no miracle materials".

Secondary prevention

In terms of secondary prevention, which means the fast identification of cases that do arise, followed by fast intervention, things get trickier. Sometimes, heat illness can be hard to spot, as the signs are not always clear. The symptoms, including fatigue, dizziness, and nausea, typically crop up first, and athletes need to feel comfortable reporting their discomfort to coaches and trainers.

Getting an athlete to speak up about discomfort or injury isn't as straightforward as you might think. Most coaches will feel that it's an asset to have a coachable athlete, someone who doesn't talk back, who listens and participates actively. But the culture of deferring to the coaches can be a problem. If you have an athlete who can't or won't speak up for themselves, that athlete could be at risk. Marty McNair wishes he'd advised his son to speak up: "I wish I'd prepared him better, told him to speak up for himself, or that in the event that you don't feel comfortable, speak up. I had always told him to listen to the coach, and be coachable. But I think as parents, we need to be clearer with our kids about the fact that they can speak up." Coaches, too, can help create a team culture that makes it easier for athletes to disclose how they're feeling, and an environment where they are working *with* their athletes rather than lording it over them.

Once a possible case is identified, it has to be quickly diagnosed. Heatstroke is diagnosed at a body temperature above 104 Fahrenheit (40 Celsius) with central nervous system dysfunction. But because there are many possible causes of central nervous system dysfunction, it's really important to get that temperature read, which is where things get trickier. The most accurate way to measure an athlete's body temperature is to take a rectal temperature, which Stearns admits has become "the most controversial part of talking about heatstroke". Only around 60 per cent of high schools in the US have an athletic trainer on staff who would have equipment to take temperature readings, but some don't carry rectal thermometers. Stearns puts the onus on trainers to carry the emergency equipment, have it near the field for every training session, practice, and game, and be ready to use it. At elite competitions, from football games to marathons, there are typically medical staff on site. But for youth and recreational sport, the equipment and medical staff may not be readily available. In this case, we rely on emergency medical services to intervene.

Once a case has been diagnosed, it takes between 12 and 15 minutes in a cooling bath to reduce the body temperature to a safe range, in most cases. An athlete is considered safe for transport to hospital or ready for other interventions once their temperature has been lowered to 102 Fahrenheit (38.8 Celsius) – and again, a rectal temperature must be taken. It's not exactly a pleasant or comfortable experience, but life-threatening illnesses rarely are. Being seen in hospital is the only way to identify any complications or organ issues that arose during the heatstroke.

Tertiary prevention

Putting together best-practice guidance on heat illness prevention and treatment is a challenging task, but it's potentially the biggest hurdle on the road to developing policies and advice to address exertional heat illness in sport. Kucera emphasizes that it can be hard to produce reliable data, and to inform policy development, without good reporting. In the past, the National Center for Catastrophic Sports Injury Research would collect information about heat illness incidents through media searches. More recently, they set up a reporting platform for coaches, trainers, parents, and doctors from across the US to submit incident reports. These provide much more reliable data.

However, most cases that don't result in heatstroke, or where the athlete makes a full recovery, don't make it into this database;

they don't get reported. So it's really hard to know how many cases have been "saves" and resulted in the athlete being successfully cooled, then sent home with minimal side-effects or long-term issues. It's also impossible to know how many cases have been prevented through new policy implementation.

Climate change is delivering more heat. A person born in the 1960s might experience four major heatwaves in their lifetime. A child born in 2020 will experience 18, and that's assuming we go no higher than the 1.5 degree Celsius warming target of the Paris Agreement. Every half degree of additional global warming will double the number of extreme heatwaves in an average lifetime. Some vulnerable countries have started using the catchphrase "1.5 to stay alive", and this can be taken literally from a health and sport perspective.

For sports events, some destinations will be wiped off the map as they become unsuitable for competition in the coming decades. A study led by Dr. Kirk Smith at the University of California Berkeley, published in the *Lancet* in 2016, projected weather averages in cities across the Northern Hemisphere in 2085 in the months of July and August (the months earmarked for the Summer Olympics) to determine how many would be able to host from a climate perspective. They included only cities with populations above 600,000, reflecting the significant expectations in logistics and financing required to host an Olympic Games. Among the 645 cities included in the study, only 33 would be low-risk Olympic hosts in a warmer future. We can only imagine what that future will look like for non-elite athletes who do not have the benefits of acclimation and acclimatization and thus will be at risk if they play outdoors in the summertime at all.

CHAPTER THREE

WILD WILD(FIRE) WEST

Coach Rick Prinz woke up on November 8, 2018 and headed to his job at the local high school like any other day. He tore through an early-morning workout, then headed to Starbucks for his morning coffee with a colleague. The local news was reporting smoke in the area, and a fire in the Sierra Nevada, not far off. But that seemed normal to the coach, who had lived through countless fires in northern California. With a playoff game looming the next day, he texted his 60-player varsity roster on the football team to confirm the day's practice would go ahead.

> Don't even think about missing practice today.
> We're practicing at three o'clock. Regardless of the smoke.
> Be there. No excuses.

From the time Prinz entered the Starbucks to the time he came out, the smoke and the flames had come within sight. Slightly more unnerved, but not yet concerned, he hopped in his truck and headed back to the school. "Ya, there was smoke, but we always have smoke," he recalled. "We're dealing with smoke all the time. So I didn't think much of it."

Around 8am, as the students were arriving for class, news started spreading that an elementary school on the other side of town had caught fire. The Principal gathered the teachers in the drama room to make the call: Paradise High School would shut and everyone would be told to evacuate.

Prinz spent the next 45 minutes helping students leave, catching those arriving and redirecting them to the nearest vehicle on its way out of town. Students were told to call their parents, but not to go home. Parents arrived at the school frantic, looking for their kids, only to find out they'd been put in a car with another family or an older student to leave town. All traffic needed to be leaving the city, not clogging up the roads by driving in different directions.

The skies darkened as the smoke settled in. Prinz recalls, "The wind was blowing over 100 miles an hour, carrying embers the size of my hand." At one point, the firefighters recorded that the fire was spreading at a speed of a football field (100 yards) per second, "and it was dark as midnight out there". When he was confident the students were all out, he went home.

After rushing his family out of the house and sending them off, Prinz looked around and tried to pack. Wildfire evacuees are typically told to pack for two to three days: a couple of changes of clothes, some toiletries, the essentials. This guidance is offered because it makes the decisions easier, eliminating the stress associated with packing up a whole life's worth of material goods in a matter of minutes. So Rick did just that. He grabbed some extra clothes, a toothbrush, and a few items for his wife who was at work in a neighboring town and unable to return home. He didn't take their computers, important documents, or valuables. He didn't think that would be necessary. But as propane tanks began exploding on his street, he realized there was no time to think about it – he had to leave.

Prinz wouldn't be able to return to his home for several months, and while the structure was spared, smoke had damaged every item inside. In the interim, he and his family moved in with friends in a neighboring town, then into a small place of their own, while the notion of going "back to normal" felt more remote by the day.

For the rest of the school year, from December through June, nothing was the same. Classes resumed just before the Christmas holiday, delivered online, with the teachers based out of the Chico mall. Teachers and students were exhausted, so only the basic curriculum was supplied, no extra-curriculars. Football was temporarily disbanded. After the holidays, school operations moved into a warehouse at Chico airport, and students showed up for in-person classes. Only 13 football players showed; the rest had been too traumatized to attend school or were displaced all over California as their families sought out new and more stable places to live. Those who attended school were keen to play again, but many were in temporary housing. With the spring training season on the horizon, and knowing the players needed something to look forward to, Prinz lied to his team: "Hey guys, we're going to have a season next fall, no matter what." He knew it was a long shot.

They had 13 players, and their status in the conference (league division) was compromised because the school had lost so many students since the fire.* But with Paradise's strong record from the previous season, other coaches and athletic directors at A-level schools didn't like the notion of playing them. For these schools, a game against Paradise was a lose-lose prospect: if you beat the team that's recovering from a fire, you're kicking them while they're down, but if they're as good as they used to be, your team would be easily beaten. Eventually, Paradise High's athletic director Anne Stearns chased down eight schools that agreed to play.

Practices were initially held on a gravel driveway just outside the warehouse-turned-school, and later moved to the field at Marsh

---

*Where they were classified as an AAA school before the fire – a school sports classification system based on the number of students enrolled at the school, AAAA being the biggest schools – they now had barely enough students to put them in A, with numbers fluctuating daily. Not having a clear categorization made it tricky to find teams to play against, because it wasn't as clear cut as it once was – big schools used to compete against other big schools, small schools against other small schools.

Junior Middle School. Gear was borrowed from other schools. The players trained in the same shirts and shorts every day, having lost everything in the fire and cobbling together whatever they could from second-hand stores and friendly neighbors. There was a lot of mud at Marsh Junior, but from the field, the players could look up to the ridge where Paradise sat and imagine themselves playing at home in the fall.

Between March and August, the team came together. They started with 13 players but fielded 45 by the summer, and nearly had a full roster by the first game. Every eligible player had returned, except one who moved to New York with his family. The rest were making it work: staying on friends' couches, in trailers, living out of their cars, working multiple jobs to support themselves so they could stay close by and play football, while their families rebuilt their lives all across the state.

It's hard to explain how important Paradise football is to the people of that town. It's central to the culture, as most men either grew up in the program or watched from the sidelines. Traditions associated with the program are passed from father to son. It's respected as a rite of passage, and in a town where there isn't much to do, football is recognized as a means of keeping boys out of trouble. So, when it was announced that Paradise football would return in the fall, the town showed up in a big way to support them.

Old players rearranged their schedules to join the coaching staff on a volunteer basis. Parents of players living in town or nearby agreed to house those whose families had moved away. Local restaurants offered lunches to the players, which Prinz learned was the only reliable meal some of them received each day. Prinz and his coaching staff rallied to drive players back and forth to their new homes in neighboring towns, and became de facto counselors to these kids who were grieving their former stability and lifestyle.

The flip-side of support is pressure. As the season opener approached, the town staked its hopes on this young fragile team.

The energy was positive but overwhelming for some players.When a town is grieving, and they attach their happiness to football games, there's significant pressure to perform. Of course, Paradise would have been proud of these boys regardless of the outcome on the field, but try explaining that to a group of teenagers.And so there were injuries and anxieties that went unattended. Prinz recalls some students hiding stress fractures and sprains from coaches, while stress and over-exhaustion were ignored.An assistant coach showed up every day despite severe medical issues, exemplifying commitment but inadvertently showing the impressionable team that medical issues and healing are not as important as football. Everybody involved in the team needed the outlet of sport to manage the internal chaos stirred up by the fire, but there was a dark side to that myopic focus: some athletes suffered mentally and physically to compete in that season.

To say the team performed well would be an understatement. They went undefeated in their season, clinching a playoff berth despite having no clear spot in any conference. When I met with Prinz in his office, off the side of the men's locker room at Paradise High in the spring of 2022, he recalled the 12-month period from November 2018 to November 2019 with mixed emotions. "It was worth it," he told me, "but the stuff going on off the field was rough. Really, really rough.Very difficult. A lot of extra time went into it, so I was tired all the time." The boys who were freshmen in 2018 were graduating in 2022, and Prinz was also hanging up his whistle and retiring from coaching. Between the 2018 fire, which impacted the whole of 2019, and the COVID pandemic, which impacted everything from 2020 through 2022, the last few years of his coaching career threw him every possible curveball, and it was time to rest.

Wildfire season

While remarkable, the story of Paradise High football is not uncommon in the western states, where wildfires have rampaged through several communities in recent years. The Paradise case was extreme, as the Camp Fire* razed the whole town, but similar stories have popped up out of Redding and Big Creek in California, and Malden in Washington State. There are lessons to be gleaned at every turn in the Paradise story. So let's go back to the start.

On the morning of the Camp Fire, Prinz didn't think much of the smoke or the warnings on the news. He'd lived with these warnings through the summer and fall seasons for most of his life; fires were sparked on a regular basis. When you're used to these warnings, they become easy to ignore.

I spent the summer of 2021 in Kelowna, British Columbia, right in the heart of fire country in Canada. From May through mid-August, it didn't rain, so by mid-June the air started to smell faintly of smoke. And then the smell got stronger. Wildfire season isn't meant to start until July, but in 2021 any sense of "usual" went out the window. By July, on the bad days, you couldn't leave the house without smelling like a campfire. On the radio, we'd hear of towns being evacuated because of the fires. One town, Lytton, just a couple of hours away by car, was burned to the ground in a matter of minutes.

We watched from our morning walks and afternoon bike rides as helicopters and small planes flew over the lake in the centre of town, picking up water to drop over the flames. On our days off, we'd spend our mornings at the beach and leave by lunch because it would get so smoky that you could barely see the edge of the water. At various points throughout the summer, you could see flames or billows of smoke rising off the ridges of the nearby mountains. At night, the flames glowed.

---

*Named after Camp Creek Road, its place of origin, the Camp Fire was ignited on November 8th, 2018 and burned 153,336 acres (620 square kilometers) in Butte County, California, becoming the deadliest and most expensive natural disaster in the world in 2018.

This is what it's like to hear the warnings and see the signs of fire every day, and to learn to live with it, because what else can you do?

As someone working in the climate space, the fires were deeply unsettling to me. I understood that it shouldn't be this bad and that something was off. The firefighters were over-capacity, triaging the fires as they arose because they had already deployed every truck, plane, and person to the frontline. By midsummer, there were no resources left. So smaller fires were inspected but ultimately left to burn out on their own. In some cases, this meant they got bigger than they should have, but so long as they weren't threatening humans, they couldn't be made a priority. Reinforcements were called in from the US, but their fire resources were similarly over-extended. So the Canadian governments sought help from overseas teams. This helped, but it was expensive and couldn't feasibly become the norm.

On a semi-regular basis, my husband and I would debate whether the air was safe enough to play in. Walking was fine, but cycling or playing tennis wasn't ideal, so we rearranged our work schedules to be available to go out in the mornings, when the smoke wasn't so bad. Neither I nor my husband have asthma, but my mom does. She found the smoke hard to handle when she came to visit. Even moderate air quality is enough to compromise a person with underlying breathing problems. Public health guidance suggests those with asthma or similar air tract issues (including COVID or post-COVID syndrome, which was rising in numbers at the time) should avoid prolonged exposure to poor air quality.

But for athletes, it's worse. When exercising, the body takes in 10 to 20 times more air than a body at rest, and most of that air volume comes through the mouth, bypassing the natural filtration systems in the nose. Consequently, a person exercising can take in much larger volumes of polluted air, and the molecules of pollution can more easily reach the lungs.

There's a prevailing attitude in the sports community that favors toughness. You'll often hear coaches encouraging players to "play through the pain" or share stories of how "in my day, we would've played without complaining". And yes, being tough and having grit is important. But often, too often, this mentality is taken too far.

The culture of toughness in many sporting environments, according to Dr. Josephine Perry, a sport psychologist, makes it hard for athletes to speak up when they're hurt or uncomfortable. "Being uncomfortable can almost be a badge of honor, and it's celebrated," she explained. "There are certain sports in particular where the culture of toughness borders on abusive, or even crosses that line, but it is expected that as an athlete, you shut up and suffer. And if you speak out, it's viewed as either annoying or unappreciative."

She calls the elite athletes she works with VIPs: very intelligent perfectionists. These are athletes who have been conditioned over time to play through pain by coaches and trainers, and by parents who are well intentioned when they tell their kids to listen to the coaches, thinking that's the best thing. These athletes develop a perfectionism around their sport that can be so myopically focused on results and outcomes that they'll let their amygdala hijack their brain and suppress the body's warnings that something might be wrong.

This can be especially complicated in team environments, where athletes tend to compare themselves to their teammates, who may also be suppressing pain. If their peers can handle it, they should be able to handle it too, or so the thinking goes. And if everybody is playing in the same conditions, it can be especially hard for an athlete to express discomfort, knowing the other athletes are pushing through.

This doesn't just apply to what we normally think of as athlete injury, but to environmental conditions like extreme heat or poor air quality, too. The problem is that it's not often recognized by coaches as a threat to athlete health, and athletes won't speak out if they feel unsafe, so it doesn't stop play. Coach Prinz was ready to practice through the smoke that day, and the athletes were ready to play along.

The impacts of playing in poor air quality are not just linked to health. Breathing bad air can impact performance. A 2018 study from the University of Ottawa and the University of Maryland was the first to empirically show that baseball umpires perform poorly when air quality is bad. A flurry of research over the last couple of years has consistently found the same: athletes make worse decisions when air quality is poor, and referees are similarly compromised. But the effects are subtle, and so they can be easily ignored or passed off as fatigue or just a bad game.

The causes of wildfires

When the fires reached the edge of Paradise, there was a mad scramble to get out. Cars sat in bumper-to-bumper traffic on the one road out of town, as the forest on both sides of the road was engulfed in flames. Fires are spreading faster than they did in decades past, making them more dangerous now than before.

The ways fires start, spread, and stop has been highly debated among the forest management community. Wildfires start for a range of reasons: sparks off an electric line can reach a nearby tree branch; campfires that aren't put out correctly can spread; there have even been some fires started by metal mountain-bike pedals scraping against roots or rocks and sparking a small flame. Usually, these human-caused fires are accidents, and they are happening at much the same rate as they always have. But some fires are sparked by lightning, which is a far more volatile and less predictable risk factor.

In August 2020, a blitz of lightning storms in central and northern California sparked more than 15,000 strikes over a few days, leading to some 600 fires and burning upwards of 2 million acres. Simultaneous lightning-sparked fires destroyed thousands of homes and buildings and claimed the lives of seven people. These are the ones most forest firefighters worry about. Lightning will strike, and char and smolder in a small area, then become an unseen blaze that can burn for days along the forest floor before it is detected. By then, these fires can be unwieldy and fast-moving. According to wildfire data published by the US Forest Service, lightning is responsible for more than 40 per cent of fires in the US West between 1992 and 2015. However, the lightning fires account for 70 per cent of the burned area, owing to their slow detection. Nationwide, an average of 11 per cent of fires were naturally ignited between 2017 and 2021. But as climate change worsens, one study by researchers at Cal Berkeley and SUNY (State University of New York) Albany suggests we could see up to 12 per cent more lightning fires for every degree of warming. In short, the number of fires – and specifically of lightning-caused fires, which sneak up on firefighting services and get out of control quickly – could rise significantly.

Climate change is also setting the stage for more devastating spread of fires. Since 2000, wildfires have burned an annual average of 7 million acres in the US, up from 3.3 million acres in the 1990s. Once a fire is ignited, a range of factors will influence whether it stays small and burns out, or spreads. Since we can't banish all sources of ignition, such as electrical lines and human activity, and it's impossible to manage winds, wildland firefighters concern themselves with managing the fuel – that is, the brush, wood, and grasses that are primed for easy burns – along with monitoring the wind speed and direction, average temperatures, and moisture levels to more clearly trace the likely direction of spread once a fire is ignited.

In recent years, droughts in the US West and other fire-prone regions have made the brush and forest floor very dry – perfect fuel. Before white settlers arrived in North America, controlled burns were routinely used by Indigenous people to manage the forest and ensure the fuel was kept at manageable levels. In a controlled burn, sometimes called a prescribed burn, the brush at the base of the forest is thinned out with carefully managed fire, which helps the mature trees grow better and removes the easily lit brush from the path of future wildfires. Controlled burns have the added benefit of managing pests and disease, and creating open space for some animals and plants that thrive in grasslands or with a little more space. However, most US states have a standing policy not to perform controlled burns, or to perform very few, despite their proven efficacy in reducing wildfire risk. Why? Because the process can be costly and complicated, and the benefits of controlled burns can be hard to measure, so they are underestimated. More pointedly, though, controlled burns don't sit well with the Western psyche.

In a 2016 research article titled "Fire in the Mind", Professor Stephen J. Pyne, specialist in the history of fire, explained the role of the Enlightenment in relegating controlled burns to the past: "The Enlightenment established fire as the dividing line between the primitive and the progressive. Primitive farmers and herders used fire; progressive agriculture found alternative ways to fertilize and fumigate." Over time, forest management was turned over to foresters, who notoriously disliked and condemned fire. This was consequential, as these foresters knew fire only as a human project, not a natural process, and fire practices were all but eliminated from land management. As lands that typically underwent seasonal routines of wetting and drying felt the impact of no burns, unruly fuel stocks built up.

Controlled burns are also not a silver bullet. Even where forests are thinned by controlled burns or other methods, they grow back every five to ten years, and land managers can only be in so many places at once to check on their growth. Forest management is an ongoing task.

Matt Reischman is the Deputy Resource Director for CalFire, the state agency responsible for overseeing fire response in California. In his role, Reischman oversees forestry resources, including preventative efforts to reduce fire risks, and the firefighting force. He told me that when he started with CalFire, the goal was to keep most fires to 10 acres or less. At the time, he explained, "we just didn't see the catastrophic fires and the rate of fire spread we're seeing now". The Carr Fire in 2018, in particular, marked a turning point for Reischman: "it created its own fire tornado, and I'd never heard of anything like that. It was essentially a tornado that was melting cars to the ground wherever it touched down."

Ignited by sparks coming off a flat tire, the Carr Fire burned more than 90,000 hectares, killed three firefighters and five civilians, and became the sixth most destructive fire in California history. The fire tornado it spurred, the first known fire tornado in California, was roughly 1,000 feet (300 meters) in diameter, burned at 2,700 degrees Fahrenheit (1,482 Celsius) – double the typical wildfire temperature – and was powered by winds circling at 135 to 165 miles per hour. The flames licked the side of the tornado, rising 400 feet (120 meters) in the air. The fire tornado essentially created its own weather system and proved absolutely terrifying for anyone nearby. The Carr Fire would go on to burn nearly 1,000 homes in Redding, California in July to August 2018, while 17 other fires burned across California. All of CalFire's resources were deployed to fight flames in every corner of the state. This was mere weeks before the Camp Fire (the one that flattened Paradise) broke out.

CalFire, like most fire services across the western states and British Columbia, makes management decisions by taking a running five-year average of fire data. The most recent data shows that the number of fires hasn't increased all that much. Rather, what we're seeing is more aggressive burns. Heavy winds and low moisture levels are leaving dense, dry forests primed to blaze. As a result, the conditions that lend themselves to catastrophic fires exist almost all year round. As Reischman explained, "Historically we were not a year-round fire department, we were staffed between May and October, and the rest of the year we're on minimal staffing. But about ten years ago we stopped saying we've got a fire season, and there's differing levels of preparedness. We'll have active fires year round all over the state. And our staffing has ramped up considerably."

CalFire sends up routine flights to capture infrared images of the state. They use these to identify new smoke and fires, and to monitor firefighting efforts so resources can be distributed efficiently. Summer 2018 was particularly scary. But in 2021, the infrared images got even scarier. "Fires were everywhere."

Triage becomes an important tool when there are too many fires active at the same time. For those familiar with medical triage in an emergency room, fire triage operates in a similar way. Quick assessments are conducted to decide where resources are needed, and typically they go to the most aggressive fires and the ones most likely to harm human life or critical infrastructure.

With the most destructive fires on record happening in the last decade, the fire budget in California has swelled from $800 million (£630 million) in 2005/06 to $3.7 billion (£2.9 billion) in 2021/22. According to Reischman, "We're still out there working with shovels and rakes and water guns, and basic tools that are actually quite primitive. And so we're fighting with nature in a way that we can't ever win." The state can deploy teams of foresters to do fire-prevention work to stop fires or slow fires ahead of them breaking out, but the reality is that the current forest conditions were created over hundreds of years of forest suppression.

Donald Trump's misplaced 2018 comment about raking the forest to solve fires was incorrect – it's far more complicated than that – but the notion that we need to take better care of the forest floor was right on target. We can't stop the fires from starting, but we can slow them down and put them out faster with more advance planning.

Not a "natural" disaster

On a research trip that took me from Southern California, through the Bay Area, Lake Tahoe, Northern California, and into

Oregon, I took a turning off Highway 5 to meet with Andrew Phelps, the Director of Emergency Management for the State of Oregon. Phelps's job has evolved in recent years from one with obvious seasonal change, to one where fire and heat are perpetual challenges.

"I've put a moratorium on the word 'unprecedented'," he told me, "and I've stripped the term 'natural disaster' from the lexicon." It's becoming clear to him that the language used by state officials to communicate with the media and the public on risks is increasingly important. "Terms like 'natural disaster' make it seem like we had no role to play here. But we did. It's important for people to know that the extreme weather or hazard is linked to climate change, which is linked to our choices. But also, the outcomes – the death tolls and the property damage – much of that can be avoided if we made better decisions about prevention and took care to protect those most at risk. They're not at risk by accident, they're at risk by design."

Over nearly a century, urban developers in the western US have eschewed warnings of flood zones, fire-prone areas, and heat-island effects, to build what Professor Stephanie Pincetl at UCLA calls "sprawling surburbia in every direction". New suburbs go up in response to calls for more housing, which Pincetl admits is necessary, but the sprawl is unsustainable. Over the course of her research career studying urban planning in California and the US West, she has demonstrated that suburbs are not only anti-social, in that they separate people from each other and exacerbate social inequities, but they're also environmentally unsafe in some cases, popping up in canyons, on the edges of fire-prone brush, and in places where nature should have been left undisturbed. People who live in these places may not know that the area is prone to fire; developers and real estate agents are careful to leave that bit of information out.

These high-risk areas are referred to by researchers as the wildland-urban-interface, or WUI. In past generations, these areas have been managed with building rules and fire codes, clear rules around how to live near fire areas. More recently, it's become clear that the main causes of fires in WUI areas are blizzards of sparks from an advancing fire that land on roofs or other built infrastructure and burn the structure itself or the flammable material nearby, like shrubs, gardens, trees, and grasses around the structure. This "home ignition zone", as it's called, is the critical point of human vulnerability when it comes to wildfires.

In Colorado, the WUI is especially at risk as grasslands have proved particularly flammable in recent years. And because grasses grow quickly, dry out quickly, and burn quickly when lit, they can be unwieldy to manage. Already, grass fires have threatened hundreds of homes across the state, including during seasons of the year that would have previously been considered "safe", like winter. In Colorado, authorities estimate that nearly 3 million people live in wildfire-prone areas.

So how can we learn the difference between a small, distant, quotidian level of fire risk and an immediate, urgent threat?

A better warning system

Prevention and awareness-raising come in many forms. In some cases, early warning systems could save lives. Air quality monitors are a similarly pragmatic tool for early warnings on wildfires and other air pollution issues.

Jamie Breslin is the Associate Athletic Director at Stanford University. In recent years, his team has faced several fire-related air quality warnings that have forced cancelations and relocations for a range of sports activities. "It's a major problem for us, and it's on the agenda for our operations team to think about every year," he explained. In some cases, games have been moved to escape the smoke: in 2018, during the Camp Fire (the same one that razed Paradise to the ground), Stanford was hosting the NCAA Elite 8 women's soccer tournament, but the air quality was so poor, owing to the fires, that the athletic department determined it was too unsafe to play.

The risks of canceling a game were huge; the way the rules were written for that particular NCAA tournament at the time, if a team cancels, they forfeit, and if a game was called off, it was unclear when it could be rescheduled in the short timeframe of the tournament. Stanford didn't want their athletes or other teams to miss out on an opportunity to move into the Final Four, so they called on their network for help and found a high school in Salinas that offered up its field. The athletes and coaches on both teams, along with officials, media, and spectators, had to be bussed two hours down to the alternate venue. It was a logistical nightmare, but they pulled it off.

More than once, Stanford football games, the athletic department's most lucrative revenue generator, were faced with tough go or no-go decisions linked to fluctuating air quality. In the absence of clear conference-wide policies governing game delays and cancelations due to air quality issues, Stanford set up their own.

Leveraging the guidance of sports medicine specialists and air quality experts on their campus, the athletic department devised a plan for responding to air quality warnings that puts their athletes' and coaches' health and wellbeing first. One of the most common ways to measure air quality is by using the Pollution Standard Index, developed by the US Environmental Protection Agency in the 1990s, which collapses several pollutants, such as fine particulate matter (pm2.5), particulate matter (pm10), sulphur dioxide, ozone, nitrous oxide, and others into one standard reading, commonly called the Air Quality Index (AQI). All outdoor sports facilities on campus have been outfitted with Perry Weather air quality monitoring systems, and if the AQI reading creeps above 150, the trainers for the teams receive an alert on their phones to take precautions by reducing the length of outdoor practices or to move indoors. At an AQI of 250, all practices and games are called off, including those held indoors.

While conservative, this approach mirrors best practice across the Bay Area. Over at Oakland Athletics (the Oakland A's), a professional baseball team, president Dave Kaval explained that his team adheres to similarly strict air quality policies when it comes to practices, and they have also installed air quality monitors. The policies governing games are determined by the league, not the team, but Kaval and the management at the A's can make decisions about practices. In fact, many sports entities in the Bay Area have appointed representatives to a joint taskforce aimed at sharing ideas, solutions, and best practices for air quality and wildfire response.

Indoor sports are not exempt from air quality challenges arising from smoke and wildfires. Seventy miles inland from the Bay Area, the Golden 1 Center, home of the Sacramento Kings in the NBA, has had smoke in the building on more than one occasion. During the Camp Fire in 2018, smoke was visible on the court, making visibility hazy and causing athletes concern. LeBron James told reporters, "you can smell it", ahead of the game, and by the end, despite scoring 25 points in 31 minutes on the court, he told the press the smoke "affected me a lot". Lakers center JaVale McGee, who suffers from asthma, also reported symptoms after playing that game. "My stomach was hurting like I was hungry or something but it was from the smoke for sure," adding, "We still won, though, so it's all good."

At the time of writing, no professional league has yet adopted any hard-and-fast policies around air quality, though Major League Baseball, Major League Soccer, and the National Women's Soccer League have begun monitoring conditions at all games and matches and will not run matches when the AQI is above the safe threshold (typically defined as anything higher than 150 on a scale of 0 to 500, where anything above 250 is considered hazardous). This unofficial policy was upheld in the summer of 2020, when smoke forced cancelations of several games in all three leagues due to catastrophic blazes that burned about 4 million acres (roughly the size of Connecticut) across the US West.

If Coach Prinz had been aware of the risks associated with bad air quality, and if there had been a school policy in place to govern responses to smoke warnings, and if they had access to up-to-date air quality readings (I know, this is a lot of ifs), perhaps he would have moved practice inside or made other plans for his team that day.

The way league rules are currently written in most elite and professional sports around the world, officials are reluctant to cancel sporting fixtures. Cancelations can upset the schedule, provide an undue advantage or disadvantage to certain teams depending on who was impacted by the cancelation, and prove a logistical nightmare to rearrange. Coach Prinz had a playoff game the next day, and knew that canceling practice could reduce his team's chances of winning. In sport, the schedule rules all. But with a long and growing list of fixtures rescheduled or canceled due to smoke and wildfire, you'd think the sports community would start building in contingency plans.

This is a challenge Kaval has considered carefully in recent years, as his team has faced several stretches of poor air quality due to nearby fires. Kaval told me that some professional team owners have proposed proactive solutions: clear policies on air quality that apply league-wide, with league-wide response plans to govern how games get rescheduled in case of cancelations. The response plans should also consider insurance and assistance for sport programs that are impacted by disasters, whether it be fire or hurricane or something else, to recover lost funds from the cancelation, and to rebuild if there is any damage. But it can be challenging to get all the owners in a league to agree, which is what it would take to get these kinds of solutions off the ground. For one thing, some owners are operating in geographic regions that are not impacted by smoke and fire, so it can be easy to chalk it up to "someone else's problem". For others, it's still too much to ask for an acknowledgment that climate change exists, let alone ask for money to address its impacts. And for many owners, there's a fear of stirring the pot with partners and sponsors: best to just keep the status quo and not admit to any weaknesses that could undermine confidence in the season. For international competition schedules, it can be even more complicated.

The long-term effects

Fires are destructive. They destroy homes, businesses, entire communities. The trauma from living through fire can be intense. Academic studies have established clear links between the trauma of living through a wildfire event and long-term rates of anxiety, post-traumatic stress disorder (PTSD), and depression. A team of researchers led by Gillian Grennan out of University of California San Diego have examined the long-term mental health and cognitive function impacts of the Camp Fire victims in 2018 and found that more individuals who were directly exposed to the fire reported having experienced trauma (66.7 per cent) compared to a control group that was not exposed to the fire (0 per cent). Several months after the initial survey, those who were directly exposed to the fire also showed cognitive deficits on a digital test administered in the lab, and reported higher levels of anxiety and depression symptoms on a self-report survey. For athletes, coaches, and their families who were directly impacted by fire, the long-term mental health impacts cannot be underestimated.

Sleeping can also be challenging after a wildfire. Not only are some people physically displaced from their homes, which can affect sleep, but the trauma from the fire is enough to keep some people up at night for several months, sometimes for several years. A 2021 meta-analysis by Australian researchers, published in the *International Journal of Environmental Research and Public Health*, demonstrated the prevalence of sleep disturbances among wildfire survivors: 60 per cent of survivors experience some degree of insomnia, while 30–45 per cent of sampled survivors report having nightmares. These numbers are staggeringly high, and will have spillover effects into all elements of the person's life, including sport.

In the period following a wildfire, given the prevalence of mental health challenges and sleep disturbances among survivors, it's critical not to put too much pressure on sports. After the Camp Fire, Coach Prinz experienced some insomnia, but pushed through it. While sport can feel like a "return to normal" and studies have shown the efficacy of physical activity for managing some mental illnesses, it's also easy to go overboard. A slow and steady return is advised, ensuring athletes, coaches, and everybody else involved are supported by their family and community and not pushed to perform too soon.

As climate change worsens, and wildfires continue to blaze near and in communities, the sports community ought to prepare to support survivors. Proactive actions, like installing air quality monitors and adopting air quality policies governing team practices, are a good start and can be undertaken by individual athletics departments or professional franchises simply and quickly.

But at a league level, whether it's high school football or professional baseball, systemic changes are also needed: ways of managing mass cancelations and rescheduling, compensating teams for losses due to cancelations or damages to infrastructure, providing comprehensive healthcare – including mental health care – to individuals in the sport system who are impacted by disasters, and adopting more compassion in the way return-to-play is handled.

Fire is unpredictable. Our responses don't have to be.

CHAPTER FOUR
EVERY BREATH YOU TAKE

Smoke from wildfires is not the only cause of air pollution. Other issues like emissions from manufacturing, chemical plants, fossil fuel extraction, and the transport sector have haunted cities – and people playing sports – all over the world. These sources of air pollution are also major sources of greenhouse gas emissions.

In other words: air pollution and the climate crisis are tightly linked. Addressing air pollution is part of addressing climate change. Solve one and you make progress on the other.

Air pollution presents a considerable health risk. The World Health Organization (WHO) estimates that air pollution is associated with more than 4 million premature deaths annually. While writing this book, I lived in London, England, where no fewer than 20 air pollution warnings were issued between September 2021 and August 2023. These warnings are not usually due to fire or smoke, despite the city's nickname "the Big Smoke". Air pollution in London, like most urban areas in the world, comes from road transport, home heating systems, and manufacturing.

While this may be surprising, there are also no uniform guidelines for air quality. In fact, most countries have their own metrics and thresholds for what is considered safe and what is considered hazardous.

To give an idea of what "normal" looks like, I use the Pollution Standard Index, also commonly called the Air Quality Index (AQI), described in the previous chapter. In this system, the parts per million of pollutants such as ozone, nitrous oxide, sulphur dioxide, fine particulate matter, and particulate matter, are calculated and categorized as follows:

- Good (AQI < 50)
- Moderate (AQI 51–100)
- Unhealthy for Sensitive Groups (AQI 101–150)
- Unhealthy (AQI 151–200)
- Very Unhealthy (AQI 201–300)
- Hazardous (AQI > 300)

In most cases, you would hope to live and exercise in environments with good air, or occasionally moderate air. But living in moderate air – or worse – for extended periods of time can be detrimental to your health, even if you're just sitting at home.

The WHO has repeatedly upheld the AQI classifications as the global standard for measuring air quality for health purposes, including for sports. And yet, due to national differences in how monitoring is done, the AQI is not always used. In some countries with higher air pollution levels, and thus higher AQI readings, such as China and parts of the Middle East, the governments have set their own thresholds for what is considered "good" and "bad", often exceeding the AQI classifications. This makes it tricky for sport, which is as international a sector as they come. For athletes traveling around the world, interpreting local air quality readings is a challenge, because they differ from place to place. Given the harsh impacts of air pollution on athletes and people who are exercising, discussed in the last chapter, sport has a fraught relationship with air pollution.

Professional athletes

When most people think of the Beijing 2008 Olympic Games, one of the first things they remember is the smog. Newscasters around the world shared images of a dark, heavy smog sitting over the city in the months leading up to the Games. A friend of mine, Emilie Fournel, a three-time Olympian in kayaking, was in Beijing ahead of the Games for training camps, and recalls looking up one day after heavy rainfall and noticing that the body of water she was training on was bordered by a mountain. She'd had no idea. She'd never seen it, despite training there every day for a couple of weeks. It was always just too smoggy to see across the water. Other athletes reported other air pollution-related challenges in the lead-up to Beijing: itchy eyes, coughing, a funny lingering smell.

With Beijing experiencing some of the world's highest emissions of sulphur dioxide, soot, and other pollutants, officials took drastic measures to clean up its air ahead of the event. First, they shut down cement factories and chemical manufacturers in and near Beijing for several weeks before, during, and after the Games. They also temporarily closed 56 power plants, most of which were coal-powered. When that wasn't enough, they banned 300,000 high-polluting vehicles from driving during the Games and halved the 3.3 million cars on the road by installing a license-plate system wherein the numbers on your plate determined which days of the week you were permitted to drive. When all of that still proved insufficient, officials stretched the factory shutdowns and driving restrictions to Tianjin, a city 130km south-east of Beijing, with a population of 9 million at the time.

One study by the US National Bureau of Economic Research calculated these measures cost the city some $10 billion (£7.9 billion), and achieved air pollution reductions of 29.6 per cent during the Games compared to the same time in the previous year. This was considered a success, and the largest air pollution clean-up experiment in history, with widespread benefits. One study published in the journal *Science of the Total Environment* showed fewer outpatient visits for asthma among adults in Beijing during the months of the Games. Locals also benefited from a weakened heat island effect: a study led by geographer Dr. Bo

Yang at San José State University monitored satellite thermal observations on the surface temperature across Beijing and noted a marked decrease of 1.5–2.4 degrees Celsius (2.7–4.3 degrees Fahrenheit) in the city.

Since the 2008 Olympics, however, Beijing has struggled to keep its air pollution under control, and it peaked in 2013. For several days in January that year, air quality readings across 30 cities in China were off the charts, reaching levels 25 times higher than what is considered safe by the WHO. Popular media dubbed this event the "airpocalypse". A key talking point in the media was the stoppage or relocation of all school sports and youth sports to indoor locations. In one particularly news-worthy case, the International School of Beijing spent $5.7 million (£4.5 million) on a dome for their athletic field to ensure athletes could breathe. But that kind of money is not available to community sport programming and publicly run schools, where the vast majority of kids play.

A year later, the Beijing Marathon was run in very polluted conditions. On October 19, 2014, the smog was visible, with 344 micrograms of particulate matter per cubic meter of air polluting the skies, an amount considered "hazardous", the worst level of pollution, by the WHO. Public warnings were issued to encourage asthmatics, the elderly, and children to stay indoors. The Beijing Municipal Environmental Monitoring Center suggested all outdoor physical activity be avoided. And yet the marathon went ahead.

Along the route, officials distributed 140,000 sponges to help runners clean their skin of pollution particles. Athletes donned masks and swim goggles in some cases. As runners approached the start line, video footage shows many of them looking up to the skies, uncertain whether to take part. Images of athletes' N95 masks taken before and after the race turn from bleached white to brown, but this was still the best available form of protection for an athlete's lungs, according to doctors. N95 respirators, which filter airborne particles, do offer some protection against small particulate matter pollution, but they are not a fail-safe solution.

Delhi, India's capital city, has faced similar challenges. The city's annual half-marathon falls in the middle of autumn, the harvest season, when farmers in the nearby northern state of Punjab burn their pastures, despite this practice now being illegal. So the air quality on race weekend has been plagued with pollution from pasture fires for several consecutive years.

In November 2017, 35,000 runners were registered to participate in the Airtel Delhi Half Marathon. Ahead of race day, air pollution rose to eight times the WHO-recommended maximums. Local schools closed, masks sold out in supermarkets, and residents were advised by local officials to stay indoors. Given the conditions, the Indian Medical Association asked the Delhi High Court to intervene and postpone the half-marathon, but the court declined, instead instructing race organizers to take proper precautions. Bharti Airtel, the title sponsor for the Delhi Half Marathon, joined the chorus and demanded that organizers address concerns of air pollution, threatening to discontinue their ten-year contract.

Organizers issued a statement defending the decision to hold the event, posted to their website in the week before race day. With $275,000 (£216,000) in prize money up for grabs, a World Athletics Gold Label, and a deep field of competitive runners who had traveled to Delhi from around the world, alongside thousands of local runners, the race director insisted the race would go ahead. Too much effort had gone into preparing for this event.

"The Airtel Delhi Half Marathon is inherently a panacea for the pollution issue that is plaguing Delhi," the statement read, and continued "Race day keeps cars off the designated 21km of the city." The company also announced they would spray the course with salt water to keep dust levels down. Nature helped, too, with a light drizzle of rain the day before that alleviated some of the worst toxins in the air. The race went ahead as planned.

The following year, organizers moved the date to October, hoping to find better air quality away from November's Diwali festival, when firecrackers are set off around the city. But as race day crept closer, air pollution in the city turned hazardous once again, proving Diwali was not the primary driver of pollution. So the organizers went further with their interventions, once again dousing the course with water vapour lined with ecologically safe chemical reactants to reduce dust, and used ultra-high frequency (UHF) radio waves to clear particulate matter from the air. UHF is an experimental technique the organizers hoped could improve the city's notorious air quality, and for the most part, it worked.

Eager to get a story on air pollution, several members of the media showed up to the half-marathon that Sunday morning hoping to interview athletes and organizers about the air quality. Instead, they got a good news story: enough pollution was cleared from the air that the skies looked mostly clear through a slight haze, the sun was visible in the sky, and organizers reported zero pollution-related incidents among the 35,000 runners.

Since 2018, the same methods have been applied annually to ensure good conditions on the course at the Delhi half-marathon. And to some extent, the organizer's point about the race bringing about better air quality in the city is correct. Fewer cars on the roads means less pollution – a key contributor to air pollution in most cities. At the London Marathon in 2018, researchers from King's College London studied the readings of air quality monitors along the race route and measured an 89 per cent drop in the city's air pollution on race day, owing to the lack of vehicle traffic. This finding inspired Global Action Plan, a London charity, to kick off Clean Air Day, the UK's largest anti-air pollution campaign.

Monitoring air quality along racecourses and at sports sites has become an important line of inquiry for healthcare providers and sports officials. Dr. Paolo Emilio Adami at World Athletics admits that there's still a lot that's unknown about air quality and its impacts on sports, particularly when it comes to long-term effects and what levels of pollution will cause performance and health challenges. However, his team is working with the hypothesis that "by being exposed to air pollution in moderate doses, the upper airways can get used to it". The idea is that the more you're exposed to air pollution, the more effective your body will become at trapping those pollutants in the airways and blocking them from entering the bloodstream. But of course, the act of "training" your body to handle pollution could come at the cost of health issues down the line. Exposing the body to additional pollution is not a good idea, because we know particulate matter can cause issues in the lungs; but the precise amount of pollution exposure a person can tolerate is not known and it's not a number anybody wants to discover the hard way. So Adami and his team have set out to unpack the impacts of varying levels of air pollution on runners through observational studies, both at the elite levels and at the recreational level among runners who participate in road races.

Monitoring might sound straightforward: just measure the amount of each particulate matter in the air, and crank out a result. Or simpler still, check online. But when each country has a different air quality index that provides thresholds for what is considered safe and what is not, it can be challenging to interpret the results.

Miguel Esperanza is an air quality specialist who has advised several organizations ranging from port authorities, to city governments, to World Athletics. Working with the company Kunak Technologies, Esperanza helped World Athletics to develop its own air quality monitoring systems and set them up at stadiums and racecourses all over the world.

The initial plans were ambitious: 1,000 monitoring stations in 1,000 cities, because people run everywhere and it's important to have local readings in as many parts of the world as possible. But this proved challenging. First, they had to figure out where the air quality monitors would be installed.

They started with a pilot project of five air quality monitors placed in city stadiums in Addis Ababa, Monaco, Sydney, Yokohama, and Mexico City. The monitors proved helpful for measuring air quality at track meets, but on most days, those spaces are empty, and not exactly helpful in providing a reading that will reflect the conditions on the roadways where most people go on their daily training runs. After all, air quality readings can differ from one spot to another, even if they're just a few feet apart. The air quality along the roadway is not the same as the air quality inside the stadium, separated from the street by the walls.

So Adami, Esperanza, and their team pivoted, focusing instead on major events. World Athletics hosts and endorses several major competitions each year, from road races, to world championship events, to the Wanda Diamond League series, and regional competitions on every continent except Antarctica. For regional and labeled races, wherever possible, air quality monitors are installed at least six weeks out, to ensure sufficient background data is available to contextualize whatever they observe on the day of the competition. For the bigger events, monitors are installed six months ahead, and for the World Championship in Eugene, Oregon in 2021, they were in place more than a year out.

Outside of stadiums and indoor events, the researchers had to get more creative. For road races, they started by mounting air quality monitors onto the back of the lead cars. But when they realized the monitor was too close to the tailpipe and picking up all those pollutants, they had to change tack and thought of bikes. But this required new technology, because most air quality monitors on the market at the time were too big to be mounted onto such a small frame. Kunak stepped in, and created a monitor the size of a shoebox. It was successfully piloted at road races in 2019 and rolled out as best-practice monitoring by the 2020 race season. Now you can find a cyclist with an air quality monitor riding just ahead of the elite women in a World Athletics road race. Since the elite women are slightly behind the leading men, but ahead of the big crowds of the general public, it was deemed the safest place to have the bike on the course, and the least disruptive.

At the end of each race, Adami and his team can produce a map of the course with air quality readings at each point along the route. Sometimes they come up with surprising results that city officials weren't aware of, based on factors like a new factory that recently popped up and is producing significant pollutants, or a roadway that has more trucks than small cars and thus is worse. The data from each race is also shared with race organizers so decisions can be made to improve the course in subsequent years, as appropriate.

The added value this air quality monitoring project brings to race organizers is hard to overstate. Organizers typically have little information about the environmental conditions along their route, unless they're in close contact with the environmental department at the city or a similar office where experts may be able to help them translate air quality readings into something useful.

Some race organizers, however, don't want to know. In some races, the traditions of the course are so strongly entrenched that it's hard to get them to change. It can also introduce all sorts of challenges if you're trying to get new permits to shut down different roads from usual.

Assessing the impacts of environmental hazards like air quality on performance in sport is complicated, because there are so many factors and it can be hard to find a big enough dataset that will allow the researchers to tease out all the other factors at play, like fatigue, pre-existing conditions, and the athletes' level of performance to begin with. Road races, like 10km runs and marathons, are some of the best environments in which to measure air quality impacts on performance, because of the high volume of finishers. This is an understudied topic; we know that air pollution is bad for the lungs and can aggravate existing breathing problems, but a relatively small body of literature exists on performance impacts.

A study run by Dr. Marika Cusick at Stanford found another large group of athletes to study: college track athletes from 46 universities across the US over a full season. Her research team found that training in poor air quality plays a role in race-day performance: those who trained in poor air conditions characterized by elevated ozone (smog) or fine particulate matter (pm2.5) had slower race times. The effect was realized even when the poor air quality was just "moderate" – it didn't even have to be "poor" or as bad as "hazardous" on the AQI index to take its toll.

In 2018, Dr. James Archsmith at the University of Maryland, alongside colleagues at the University of Ottawa and the University of Manitoba, discovered another great opportunity to measure the impacts of air quality on performance: baseball umpires. Because there are so many MLB games in a season, and umpires travel all around the country to call games, it's possible to observe the same umpires in action in different geographies. Thanks to pitch tracking technology, there is also a reliable way to verify whether umpires made accurate calls.

Archsmith's study examined umpire decision-making in 12,543 games and found that just a 1-part-per-million increase in carbon monoxide causes an 11.5 per cent increase in the propensity of umpires to make incorrect calls. This equates to roughly two more incorrect calls per 100 decisions. However, increases in ozone and nitrogen oxide in the air didn't have an effect. In a sport obsessed with statistics, this is all noteworthy.

So we know air pollution is bad for performance and health, and some air quality monitoring efforts have been rolled out for athletes and sports organizations at the top of the sports food chain. Progress is being made. Still, these air monitoring efforts do little to protect and insulate the hundreds of thousands of young people playing in hazardous conditions at the lower levels.

Amateur athletes

In late 2021, I spoke with Eric Holthaus, a St. Paul, Minnesota-based meteorologist, dad of two, and author of *The Future Earth* (one of my favorite books on the climate emergency) about early warning systems. Eric had noticed that most people were relying on generic weather data to make decisions about their day: everything from what to wear, to whether they could sneak in a bike ride before the rain began, to the likelihood of black ice forming on the roads in Minnesota's brutal winters. Weather information was coming from the weather app on people's phones, or in the 60-second weather announcement on the radio. Often, this information wasn't specific enough for those working outdoors or for whom the weather could impact the course of their day: people traveling, using active transport to get to work, working the land, or playing outside. So he formed a small team and developed a text message-based weather service that offers customized weather alerts, with interpretations for the impacts of each weather event, to subscribers in every corner of North America. The service includes air quality warnings.

This is a lifesaving service, and one that's affordable for high school athletic departments, youth sports clubs, and volunteer-run organizations. But supplying this type of information without coupling it with education on the implications is unlikely to have the desired impact. Coaches, trainers, athletic directors, even parents of athletes need to understand the risks of air quality issues.

After a few bad wildfire seasons and pretty heavy smog days in Toronto, Canadian sports authorities have started to take note of air pollution. In 2022, I was invited by the country's Sport Information Resource Centre (SIRC) to help develop an air quality training course for coaches. The collaborative process, involving several members of the SIRC team, many medical doctors, representatives from Health Canada, and sports stakeholders, took only a few months but resulted in an easy-access resource for educating the sports sector on air quality issues. Hopefully, it will make people aware and keep them safe.

In the training, the dual focus is on awareness and adaptation. Simply knowing the air is bad isn't enough; coaches and athletes need to change their training and competition plans.

When I asked Adami what athletes can do to avoid ingesting polluted air, he laughed. "Not much." Runners will often turn to masks, but these are not particularly effective for blocking polluted air and can be uncomfortable to wear while exercising, as most athletes discovered during the COVID pandemic. The more effective strategy, though admittedly annoying, is to move indoors, or switch the timing of the event to early morning before people go to work, when air quality tends to be a bit cleaner due to reduced traffic and less daytime heat.

It can also be helpful to consider where a sporting activity is taking place in a city. As Adami and his team have shown in several studies, air quality differs all along city marathon courses. This means there are cleaner and dirtier parts of a city, so better and worse places to play outdoors. Avoiding major roadways, ports, airports, and factory areas can improve your chances of finding clean air, even in a polluted city. Trees and green spaces also help.

Trees, especially coniferous ones that don't lose their leaves in winter, and trees in parts of the world that keep their leaves all year round, are brilliant pollution-suckers. Trees breathe in carbon dioxide and breathe out oxygen. Pretty ideal, considering humans do the opposite. Their efficiency in reducing the heat island effect, pulling pollutants out of the air, and capturing carbon have propelled trees and reforesting projects to the top of the climate agenda. But just from a health standpoint, those who live near trees enjoy significant benefits.

A US Forest Service study published in the journal *Environmental Pollution* used computer simulations of local environmental data across the conterminous United States and revealed that trees and forests removed 17.4 million tonnes of air pollution in 2010, with human health effects valued at between $1.5 billion and $13 billion (£1.2 billion and £10 billion). If that's not a win, I don't know what is. So, if you're going to play outdoors, play in parks or near trees.

Let's stop here to discuss another key issue: environmental equity. Not every neighborhood has equal access to trees. Some neighborhoods, especially fancy ones with big homes and wealthy people, are leafy and lush. Others, typically those with low-income residents, live without trees, and miss out on the benefits trees provide both for heat alleviation and air pollution reduction. American Forests, a non-profit organization, analyzed tree cover in US cities and found that these income lines also fall commonly along racial lines: communities of color have 33 per cent less tree canopy on average than majority white communities. Achieving tree equity would require reconfiguring several neighborhoods in every city to make space for trees, along sidewalks, roadways, in gardens, and expanding existing greenspace. Once there's space, more than 500 million trees would be needed to approach equity. That's a huge gap.

It may also help to avoid playing outdoors on big game days if you live near a major stadium or sports event. Air quality monitoring efforts have revealed negative effects of sports events on air pollution. Unlike marathons, where significant roadways are closed to traffic for a portion of a day, some spectator sports actually see notable increases in local air pollution from traffic and fan celebrations. In Santiago, Chile, winter brings high concentrations of particulate matter, peaking sporadically around ten times above average levels. City officials have focused on decreasing pollution from traffic, industry, and residential heating, but those temporary peaks in air pollution weren't going away. Researchers studying this trend examined different possible sources, and discovered backyard barbecues on football (soccer) game days as the culprit. It might seem such a small thing – lighting up the barbecue to cook for your friends on game day – but when the whole city lights up at the same time, it can overwhelm the air. The study found that game day barbecuing was responsible for at least one in every four air quality alerts in the city since 2014.

Tailgating (setting up a barbecue with the car's tailgate or boot open) at football games in the US has similar effects.

Colleagues of mine, Dr. Jonathan Casper and Dr. Kyle Bunds at North Carolina State University, have been monitoring air quality in tailgating zones and the stadiums of college football teams on game days for several years. Their research has shown that grills, generators, and cars in tailgating areas cause pollution levels to spike, sometimes up to 20 times worse than the recognized levels for moderate air quality. However, while tailgating activities can stretch on for several hours ahead of the game, as fans trickle in and join the party, the highest levels of pollutants were actually observed after the game ended, due to the post-game rush of traffic.

In professional sports franchises from Seattle to London to Tokyo, staff are starting to explore options for reducing game-day traffic. This will be easier to do for stadiums and arenas that are located near public transit and city centers. For them, it's as simple as offering free public transit access with the ticket, though this needs to be negotiated with city officials and the transit authority. Where the sports facilities are located further out from the city center, there tend to be fewer transit options, so it's trickier. A friend of mine, Monica Rowand, a sustainability executive at Waste Management, once joked that sports teams should start offering party buses from park-and-ride lots around the city to their game, reducing traffic closer to the stadium. She was joking, but I'm not: that would take the edge off air pollution and reduce fan travel emissions on game day. Instead of tailgate areas with hundreds of cars, these buses could be organized to take fans to the game, just as some European football teams organize buses for supporter groups to be shuttled to away games. These keep fans together, and keep the party going (without the need for a designated driver or any unnecessary emissions) for hours before and after the game, extending the game-day experience. Win-win.

With more awareness of air pollution and its tight links to greenhouse gas emissions and climate change, it's likely we'll see more air quality efforts coming from sports organizations in the months and years to come. Already, World Athletics have launched a global campaign asking politicians to take meaningful action on air pollution. The campaign, "Every Breath Counts", was built off the back of their air quality monitoring efforts, and has been covered in national media in the UK, the US, Monaco, France, Kenya, Sweden, China, and elsewhere. People can't play if they can't breathe. It's really that simple.

CHAPTER FIVE

PLAYING ON THE EDGE

In 2016, while Donald Trump was on the presidential campaign trail promoting a border wall, his company was applying for building licenses to erect a different kind of wall. Perched on Ireland's west coast, Trump International Golf Links & Hotel has faced significant erosion over the last decade, prompting a plan to build a 28-kilometer (17.5 mile) coastal protection wall.

Trump, who has called climate change a "hoax" and "pseudoscience", clearly has a different take when it comes to his golf course. The applications for permits to build the coastal protection wall explicitly cited rising sea levels, global warming, and extreme weather as the primary justification. The first application was submitted in 2016, but denied by the Clare County Council thanks in part to the #NatureTrumpsWalls campaign, spearheaded by Save the Waves, which gathered over 100,000 signatures and 700 letters of objection in a petition against the project. Trump's company went back to the drawing board, and proposed two separate walls at either end of the beach, 630 meters and 260 meters in length respectively (690 yards and 285 yards). This proposal was successful at the Council.

Local residents had mixed views on the project. Some lauded it for protecting the golf course and thus jobs at the resort, which employs 300 people, and in the town. Others were concerned about how the barriers would impact dune habitat and questioned the efficacy of small barriers, which would only serve to send the waves further down the shore where they could wreak even more havoc on neighboring lots.

Doughmore beach, which sits along the coastline just below the golf course, is inside a nature reserve called the Carrowmore Dunes Special Area of Conservation. The spot is popular among surfers, with a gently sloping sandy beach bordered by a steep cobblestone shore called a storm ridge. Between the beach and the golf course there is a system of dunes.

Coastlines are not stagnant. They move and shift, usually slowly, over time, but sometimes very quickly, as when there's a storm with big waves. The system works like this: waves hit the shoreline, and the energy is dissipated by the beach, predominantly by the storm ridge (cobblestones) at the top of the beach. In stormier conditions, the wave will reach past the storm ridge and the energy will encroach on the dunes, causing some erosion. Basically, the waves will push the dunes back into the land a little. If there are breaks in the dunes, the water can stream between the dunes and pool on the flood plain – in this case, the golf course – on the other side. But it doesn't stay that way for long. Again, nothing is stagnant on a coastline.

When the seas are calm, water will seep back out to the ocean, leaving the dunes and beach exposed to future king tides and storms. This happens on coastlines all over the world and is totally natural. But with increasingly stormy weather and sea-level rise, this accelerated process could lead to faster and more significant changes along the water. This is an inconvenient truth for people who live along coastlines or who want to protect their golf course. In theory, some of the sediment might come back, but it could take years or decades, and it's not guaranteed. Golf courses don't want to wait that long.

The Trump project didn't immediately go ahead as it got caught up in a years-long back-and-forth battle of approvals and appeals. By 2020, An Bord Pleanála, Ireland's independent planning body that decides on appeals over planning and building, put an end to it, overruling the scaled-down proposal, and citing concerns over the efficacy of the project and the resort's long-term future.

"The board is not satisfied that the proposed development would not result in adverse effects on the physical structure, functionality and sediment supply of dune habitat within the Carrowmore Dunes special area of conservation," it said in a statement.

The saga continued in 2022 when Friends of the Irish Environment (FIE) submitted a complaint to the country's Minister of Heritage over a sand fence that had been erected on the dunes. The fence consists of several large wooden poles set into the ground, with wooden planks running horizontally across them holding a perforated plastic sheet in place. It's an eyesore and according to the FIE complaint, the fence is not in compliance with legal requirements for the dune system owners to abide by conservation objectives and maintain the natural movement of sediment and organic matter to and away from the coastline, without obstructions. FIE director Tony Lowes told the *Irish Times* that they were dealing with an organization that "didn't seem to be able to admit they had lost".

From a golf standpoint, unstable land, soggy greens, and weak shorelines can be dangerous for golfers and golf carts driving the course. When courses flood, they close, because it's unsafe to play. The loss of revenues during a closure can be hard to swallow for the management, but it's also an expensive problem to fix.

To avoid these outcomes, drainage on golf courses is usually pretty sophisticated, consisting of a primary system of lateral drains and underground piping to redirect water away from the course toward ponds or other capture sites for reuse in irrigation. A secondary drainage system of natural features is designed to direct surface water that pools on the course toward places where the primary system can take over. These features include sand and gravel slitting, where vertical channels are carved into the surface to direct the water away, ditches, and contouring. When these drainage systems are overwhelmed, because of heavy rainfall or a

storm surge, it can take extraordinary measures to get the water
off the course. Coastal golf clubs will do just about anything to
avoid that scenario.

So it came as no surprise when in 2021, another Trump golf
course adopted a similar protectionist approach. Reports of a
soil stabilization project at the company's West Palm Beach club
made local news, but this one was done more responsibly, and
was largely successful in preserving the integrity of the course
without compromising waterways on and near the property.
Erosion is natural, particularly in older courses and those along
coastlines, but the solution can't be to destroy natural features and
build grotesque walls all over the place.

Trump's controversial Aberdeenshire golf course in Scotland –
for which the company all but destroyed a protected dune system
– is also at risk of severe flooding due to climate change. A study
by the Ordnance Survey predicts the coastline next to the Trump
resort will recede by tens of meters over the next 20 to 30 years.

I contacted the Trump courses several times between 2021 and
2023 to ask about all of this, but they declined to comment, just
as they have done with journalists covering the saga.

The trouble with golf

Erosion challenges are not unique to Trump's courses, although
those are particularly ironic given his record of denying climate
change and associated sea-level rise. Dozens of other golf courses
face a similar fate.

Roughly one-sixth of Scottish golf courses are located along
coastlines. St Andrews Links, one of the oldest courses in the
world, located about an hour and a half north of Edinburgh,
loses 1–3 meters (3–10ft) per year to sea-level rise and coastal
erosion. In February 2022, I took a trip to visit St Andrews and
met with Ranald Strachan, the sustainability manager at the
site who's been hired in part to protect the links from further
damage due to sea-level rise. Ranald has lived in the area and

worked along this particular coastline for the better part of 25 years, and the ease with which he navigates the sand dunes and tall grasses reveals the joy he derives from doing this work.

St Andrews Links serves as a buffer zone between the North Sea and the city of St Andrews, home of the storied university by the same name (Prince William and Kate Middleton met there as students). Over more than 150 years of its existence, course managers have designed seven exciting and competitive courses for players on the grounds. But those courses didn't come easily: large swaths of the site had to be drained of excess water, creeks were filled in, and the courses were landscaped to suit the sport, not the natural features of the site. Consequently, when bad storms happen, some of those natural features re-emerge.

Ranald walked me through a few spots on the courses where divots and troughs flood when there's a storm or a surge; those used to be creeks, so the water naturally wants to flow back there. The course's famous Swilcan Bridge, which spans a small stream on the 18th hole, has been undermined by erosion and has had to be rebuilt several times.

In recent years, St Andrews Links has seen a few flooding events due to storm surges. In late March of 2010, Ranald remembers waking up to a wild storm that produced a surge strong enough to breach the dune system along the site's coastline. Water pooled on the course, completely overwhelming its drainage capacity. In 2013, a second heavy storm surge produced similar flooding issues.

After both events, St Andrews Links attempted to restore the dunes. So far, a range of approaches have been put to use. In some places, gabions or rock armour were built to prevent the water from rising up to the level of the course. In other spots, the solution has been to plant grasses and build sand dunes, with the same effect. Volunteers have been a big part of making this work. Since 2010, over 1,500 individual volunteers have contributed thousands of hours to removing invasive plants and restoring the dunes to ensure their long-term survival.

A range of support projects have also been undertaken by volunteers such as the "Christmas Trees – New Dunes for Old Trees" project, which uses recycled Christmas trees to reinforce the dunes, and the repurposing of 12 acres from municipal parking at the site to grazed and managed coastal grassland since 2014. Additionally, West Sands – the beach that runs along the coastline by the golf course (and famous for featuring in the movie *Chariots of Fire*) – now has a Code of Conduct for all beach visitors to ensure they don't do any damage to the dunes.

Dr. Clare Maynard is a saltwater marsh specialist based at the University of St Andrews, and the head of the Green Shores Partnership, an alliance of local land management agencies, researchers, and coastal properties like St Andrews Links. Maynard has been studying solutions to protecting the shorelines in Scotland for the better part of 30 years and has had success with some of her rewilding initiatives. Over tea in the clubhouse at St Andrews, Maynard walked me through the various approaches to coastal erosion and sea-level rise.

"There's three ways to go," she started bluntly, "you can build a wall, you can replant and rewild the dunes, or you can give the land back to the sea." I pressed her on the last option, asking what that meant. She nodded at me, "exactly what you think. We pick a part of the course and we designate that as the area that will be flooded. We sacrifice part of the course, to protect the rest of the course and the city that sits behind it."

Grim as it sounds to sacrifice part of the course to the sea, the alternatives are not great either, and serve as stop-gap solutions. A gabion, which is a cage or container filled with rocks, sand, or concrete to prevent erosion, can cost $3 million to $5 million (£2.4 million to £4 million) per kilometer, and can require rebuilding every few years to ensure it works efficiently. Rock armour, also known as riprap, which is essentially rock walls strategically placed to protect coastlines, costs a similar amount. These structures can be devastating for wildlife, cutting off important access routes for insects and small animals, and splitting the local ecosystem in half.

Dunes, on the other hand, are much healthier for the animals and plant life, but they erode quickly. A mature dune will shrink over time, and grow marram grass, which stabilizes the dune. It will reduce the amount of water that reaches the golf course, but it won't stop it altogether.

What concerns Maynard most is the accelerated pace of sea-level rise in recent decades, and how that's butting up against the British mentality. She tells me:

> What I'm doing and what Ranald is doing is a waste of time in the long run, because none of these dunes or gabions will resolve the fact that sea level is rising and coastal erosion is inevitable. I worry we've already hit the tipping point. We're there. The speed of acceleration is what scares me now. And the fact that the real solution, of giving some land back to the ocean, goes against British mentality. We're on an island. And the dominant mentality is that you should defend against the sea, you don't bring it in.

Her proposed solution is to create a floodplain, potentially sacrificing a course or a couple of holes.

> We know where the water wants to go, it's shown us as much in past storm surges and floods. So let's let it go there. If we continue restoring the dunes, and build a second wall further back from the dunes to draw the boundary on how much land we'll give back to the water when the time comes, that could be our best plan.

Strachan agrees with the notion of moving the courses inland to facilitate creating a floodplain. But he emphasizes that no plan is perfect, especially in the long run. "The Links management team wanted me to make up a coastal plan for 70 years. And I just said, I'm not going to do that. Because you can't plan for something when you really have no idea what's going to happen." While long-term planning is not possible, coastal erosion and storm surges can be reliably predicted in the near term – five to ten years. Beyond that, the focus has to be on adaptation and flexibility. "If you pull back the golf course, you've sorted it," Strachan offered. "But if we're trying to protect this iconic spot, if we're going to stay here, we've got to get these next ten years absolutely right."

The irony of St Andrews Links is that the site only continues to exist because there's been quite a lot of human intervention to protect the coastline in the first place. And now there's going to have to be more human intervention to preserve what's been done in the past. It's a never-ending loop.

When I visited in 2022, the Links were preparing for their 150th anniversary event, and a bigger conversation was beginning behind the scenes to map out possible adaptations to the site that would ensure its longevity in a climate change-impacted future. Ranald told me that he thinks we may have to consider more radical approaches: focusing on changes to the game, not the site.

> If you say to people, look, this course might change in 20 years, they'll go, yeah, that's fine. And hey, I would have thought that would be quite good, if you're really interested in golf. Because, you know, it adds to the challenge. We can't just be pandering to people who want to play the same course, so that they can sit in the pub with a pint at the end of the game and say "I'm still at scratch. I'm still great." Well, that's probably because you play the same bloody course every day! But I do get the sense that golfers are changing, I'm optimistic. You know, the club would have had no traction with these discussions ten years ago. And they are beginning to understand it now. When I talk about the environment, I'm not getting pushback anymore.

Maybe the golf world is almost ready for the advent of mobile, challenging nine-hole courses, instead of 18 holes. That kind of change would result in a course that takes up less space, with a much heavier focus on better, higher quality of design that responds to the environmental changes and can be adapted over time. Those nine holes can be shuffled in order, or moved around on the site. Another proposal would be to have the industry go back to club manufacturers and have them stop over-manufacturing clubs with heads that can drive the ball across an entire course, because it's just not sustainable. Already, some manufacturers are downgrading golf balls because they can manufacture them to go as far as you want, but they're now coming to realize that bigger and longer isn't always better in golf. Sometimes, those mid-range distances are just as challenging, if not more so. While it's impossible to say exactly what changes are coming in the sport, it's clear some will be needed.

Moving the goalposts

Other coastal sports have fewer adaptation options when faced with erosion. Tuvalu is a tiny Pacific archipelago of nine islands over 26 square kilometers (10 square miles). The country made headlines during COP26 when its foreign minister made his address from the Tuvalu coastline, standing knee-deep in water. Its beach volleyball team has been a surprise success in recent years, earning berths to the Commonwealth Games by winning the Pacific Mini-Games in June 2022. But they don't have any home courts. There's just no room on the beaches, which are thin and slope steeply. Over the years, the athletes have watched the beaches fade as waves lap the coast. Forced away from those beaches, they train at Funafuti International Airport, the only open space in the archipelago. It's a popular sports site, since it only receives two planes a week, leaving it open most of the time for football, rugby and, more recently, beach volleyball. The athletes carve out their space to train alongside other sports.

But even at the airport, conditions are inconsistent. Some days, the team shows up to find that high tides have caused waves to come on land and wash away their sand. So they've started holding little training camps on other islands, some two hours away by boat. But these aren't glamorous: they're not staying in hotel rooms, but in tents. It's simple, but their commitment to adaptation has saved their training program.

"In these guys' lifetime, if nothing changes, their homeland won't be there anymore," Tuvalu's beach volleyball coach, Australian Marty Collins, told the media, after his players Ampex Isaac and Saaga Malosa lost their opening match to England at the 2022 Commonwealth Games. There may not be a Team Tuvalu in future.

Elsewhere in the Pacific, a New Zealand sailing club had to move its clubhouse back from the coastline to avoid it falling into the sea. Mercury Bay Boating Club had long known it was in trouble. Located on a peninsula of the North Island, the storied club was instrumental in the challenge for the 1988 America's Cup. Volunteer-run and community-focused, its clubhouse has become a gathering point for the locals – not just for regattas, but also for weddings, luncheons, meetings, and other community events. The clubhouse sat on green space owned by the local council, just meters off the edge of the water.

Over several years, a local surveyor – the dad of a kid who sails at the club – has been monitoring with high accuracy the accretion and erosion after every storm event, producing detailed drawings to show how the coast has changed. Noting the lost beach area over time, the club's commodore Jonathan Kline entered into conversations a few years ago with the local council to propose coastal protections: some rock on the beach, possibly a rock wall. Previously, two rock walls had been installed along the beach after Cyclone Bola in 1988, to protect homes and other infrastructure. So there was some precedent. But the council and coastal authorities said no, citing a new nationwide policy to implement soft defenses and move toward managed retreat from impacted coastlines.

So the club started thinking about managed retreat back from the coastline, drawing up plans to pick up the clubhouse and move it backwards some 50 meters (165 feet) from the water on the same plot of land owned by the council. Fundraising for this – which would cost in the range of $200,000–$250,000 NZ (£95,000–£120,000) – began in earnest. The club thought they had a few years to fundraise and figure this out. But in 2023, things got more serious.

First, Cyclone Hale hit the coastline at the end of January 2023, taking with it 6 meters (20 feet) of coastline over two high-tide cycles. When Kline and his team arrived at the building after the storm, the beach deck was dangling over the edge of the water. With a group of volunteers, Kline got to work cutting off the deck and deploying 64 concrete blocks and 100,000 sandbags to protect what was left of the building. They were concerned that in another high tide, the whole building would go. It was a temporary fix, but it wouldn't hold up in another storm. So Kline called the moving company.

There were two options for moving the building. The first option was to apply for permissions to move the building straight to its intended permanent location at the back of the site. But these permissions could take months to acquire. The alternative option was to move the building back partway, set it down illegally in the parking lot to get it away from the water's edge, and then move it again to the permanent spot when the permissions came through. The cost difference was roughly $25,000 NZ (£12,000) – moving it twice was expensive. However, in the aftermath of Cyclone Hale, it became clear that the process to get permissions for the permanent spot would take far longer than Kline was comfortable with. What if another storm came and washed the building out to sea while they were waiting? What then?

So Kline made the tough call to spend the extra money to move the building partway back, 28 meters (nearly 100 feet) away from the coast, into the parking lot. As I write this, the building is still in that lot. It was there when Cyclone Gabrielle hit just weeks after Cyclone Hale and took another 6 meters (20 feet) off the coastline, crunching 28 meters of distance from the sea to 22 meters (70 feet). Had the building not been moved, it would have been lost. It's clear it can't stay in the current temporary location, but the approvals process to move the clubhouse to its final destination is complicated.

The total costs of this adaptation effort will reach into the range of a quarter-million dollars, about £120,000, when all is said and done. And that's cheap; many of the professional services associated with the project – the architect's fees, the urban planner doing the paperwork, the surveyors – are charging half price or working pro bono, because they all live in the community and either use the club themselves or have children involved in its sailing programs. Not all volunteer-run sports organizations would have that kind of community support and network to draw on.

Throughout the process, Kline has kept kids on the water. Sailing has continued through the summer (when the storms weren't raging), and he was able to find a shipping container to set down on the beach to store equipment closer to the shore. So, even as the building was being lifted onto a truck bed and moved backwards, the club's annual Steinlager keelboat regatta took place in the bay beyond.

The clubhouse is unlikely to find a permanent home soon. It's a complicated process that could take months if not years to resolve, with several rounds of approvals and consents needed to set the building down on the site they've identified as most appropriate. And still, even if they clear all those hurdles, there's no guarantee that future storms won't further erode the coastline and threaten the building's new location in just a few years' time.

Kline, who never expected his volunteer role as a sailing coach and club commodore to turn into a permissions nightmare and crisis management role, has advice for other sports clubs: "Start early, lean on your community, and don't wait." The fact is that if you're in a coastal location and you're noting erosion, you might think you have time, but one major weather event can run down the clock.

After Cyclone Hale, the clubhouse got national news coverage and soared to the top of the local council's priority list. But once Cyclone Gabrielle hit just weeks later, Mercury Bay Boating Club fell back to the bottom of the list because local roadways and other critical infrastructure were impacted and needed urgent repair. Sports facilities are not considered critical infrastructure.

Alexandra Rickham is a three-time Paralympian with two bronze medals in sailing and five World Championship titles, who transitioned to front office roles after her retirement from the sport. She is now the Head of Sustainability at World Sailing. In a recent conversation, I asked her about coastal erosion in sailing and she was immediately able to rattle off the names of several sailing clubs in the Caribbean and South Africa that had recently been hit by storms and had lost docks, boats, and part of the coastline. "This definitely happens," she told me, "and it can be tricky to get boats back on the water afterwards because everything is either in ruins or it's so unstable that you wouldn't want athletes on the docks or near any of the buildings that have been compromised."

Riding the waves

The other coastal sport worth mentioning here is surfing. Since the start of the pandemic, surfing has grown by 35 per cent as millions of recreationists have picked up a board to escape their lockdown boredom. Another segment of surf enthusiasts are coming into the sport as fans of Apple TV series *Make or Break*, which follows the World Surf League professional tour. What's not shown in the series, and may be hard to recognize for those new to the sport, is the challenge of eroding coastlines and surf breaks all around the world.

Dr. Jess Ponting at San Diego State University has been studying surf tourism and the impacts of climate change on the sport of surfing for decades. He has been monitoring the loss of sand on beaches in Southern California where the water is approaching the bluffs (cliffs). But he emphasizes there's more going on when the big waves roll in:

Everyone walks along the boardwalk and sees the water spray up into the air when there's a big surf and that's really fun to see.

But what you don't see under the water is that the wave is not just moving vertically, it's moving downward as well and digging up sand, pulling it back out to sea. And it's a very difficult process to get that back up onto the beach. In fact, it rarely happens, and we now have a lot of beaches eroding away. In some places, the water now comes all the way up to vertical bluffs.

The sand on beaches – both the stuff you see on shore and the sand under the water where the wave breaks – is being swept away in some places. And that's impacting the surfing opportunities even at the best of times.

Surf events are impacted too. The World Surf League (WSL) has been running events at Trestles, an iconic beach in San Clemente, California, since 2001. But in the last three years, they've noticed significant erosion forcing organizers to reduce the event's footprint by 50 per cent. The entire main structure had to be moved to the inland side of the roadway as high tides threatened the events' electrical equipment. In an email, Bob Kane, WSL's Chief Operations Officer, explained that "With the possibility of further erosion, we are evaluating alternatives to enhance the fan experience, including exploring activation opportunities in downtown San Clemente." This might bring more fans to the festivities, as a downtown location may be more accessible to some, but it definitely doesn't have the same vibe as being on the beach.

Similar sand loss has forced the WSL to move all their event infrastructure off Sunset Beach for their annual competition held on Oahu's North Shore. The event is now based out of two pre-existing residential homes nearby that house the judges, media, broadcast production facilities, and athlete areas. The move wasn't optional, it was necessary. According to Bob, "Without these adjustments it would have been impossible to successfully run the event."

With sea-level rise, we have to imagine that future surfing conditions will look more like what high-tide conditions look like now: in other words, completely unrecognizable. The Scripps

Institution of Oceanography has modeled the effects of sea-level rise on beaches and surf breaks in Southern California and has concluded that at some of the more eroded beaches, waves will become fuller, fatter, shorter, less intense, and generally not so good for surfing.

A study by Dr. Dan Reineman, coastal management researcher at California State University Channel Islands, projects that 18 per cent of California's surf spots are at risk of being lost by the year 2100. Of the remaining locations, 16 per cent could be less suitable for surfers owing to changes in the way the waves form. Only 5 per cent of the surf sites in California are expected to improve. This will be a heavy loss for surfers in California. In a survey of surfers' attachment to their beaches, Dan and his colleagues found that many regular surfers consider the beach "part of their family". For them, it would be devastating to see it go.

In Australia, North Sydney's beloved beach Narrabeen, a National Surfing Reserve since 2009, has an illustrious history of hosting surf competitions because of its idyllic waves. In recent years, the stretch of beach has made headlines for other reasons: big swells are causing massive damage to nearby houses. In 2016, a fierce storm caused so much erosion that a decision was taken by the local council to erect a sea wall 7 meters (30 feet) high. The intention was to protect the homes along the coast. It took several years to get it approved and funded, with residents funding the majority of the building costs. But the result is ghastly – the wall is 1.3 kilometers (almost a mile) long, stretching from Colloroy Beach to South Narrabeen, built from concrete, and is worsening the sand erosion problem on either side of it. It was met with outcry from the Surfrider Foundation and other community environmental groups who fear the wall has just put the future of the whole beach at risk.

Other beaches in Australia are facing similar challenges. In late 2020 through 2021, Main Beach in Byron Bay faced several storms that pulled away significant amounts of sand. At Inverloch in Victoria, the surf lifesaving club has had to retreat twice from advancing tides, as the beach has lost 33 meters (36 yards) of beachfront since 2012.

Australia's 12,000-kilometer (7,500-mile) coastline could face continuous issues with sea-level rise and coastal erosion, as the Intergovernmental Panel on Climate Change (IPCC) projects that the future rise in sea level caused by thermal expansion, melting ice sheets, and land water storage, could be between 0.43m (17in) and 0.84m (33in) by 2100, depending on how much greenhouse gas the world emits. It's unlikely the coastal defenses will all survive sea-level rise and the increasing frequency of storms and high tides that will lash at coastlines. A March 2020 study published in *Nature Climate Change* suggests that by 2100, half of the sandy beaches worldwide could be lost.

It's not just the beaches that are eroding. Surf researchers are also increasingly concerned about the impacts of sea-level rise, ocean temperature rise, and ocean acidification on coral reefs, over which many of the best waves in the world break. Jess suggests the good news is that coral reefs seem to be more adaptable than we once thought: "I look back at my lecture slides from 15 to 20 years ago, and it was like all the coral reefs in the Indian Ocean are going to be dead by 2030. That's literally one of the slides that I had. And that was the prediction that was being made in the very early 2000s. And that doesn't seem like it's going to come to pass."

It's clear that the corals and the zooxanthellae* (pronounced zoo-zan-the-lee, I had to look it up), which are in a symbiotic relationship with the corals, seem to be able to adapt to a certain extent. Yet we haven't seen the full brunt of acidification that will come with temperature rises, and that is expected to dissolve the coral. What does that mean for surfing? If we see massive coral

---

*Zooxanthellae are single-celled photosynthetic organisms that live on corals (and other invertebrates), providing the corals with up to 90 per cent of their energy needs via their photosynthesis process.

reef die-off, and then its destruction and erosion, some of the very best surf breaks in the world could disappear into a swell.

A number of sports organizations have stepped up to solve this problem. The World Surf League, for example, has a whole program dedicated to conservation and erosion mitigation at stops along their tour, including in Hawaii, Australia, El Salvador, and Brazil. The program facilitates mangrove and coral replanting efforts, water quality controls, beach clean-ups, and athlete engagement on ocean issues. Meanwhile the WSL Pure Grant Program funds non-profit organizations focused on coastal restoration and plantings, erosion mitigation, conservation, climate change adaptation, and marine plastic pollution.

Save the Waves Coalition is an international non-profit working to protect 1,000 surf ecosystems by 2030 through campaigns, effective stewardship, and protected areas. It's an ambitious goal, but as CEO Nik Strong-Cvetich explains it,

> for every beach, we build a coalition, usually some combination that includes the city, the county, local surf clubs, the Sierra UC: Club, the Surfrider Foundation, and so it's never "Save the Waves saved it", it's "that working group saved it". We're always working toward shared victories, which means the impact was shared. Some of these places are very small. We're talking about 2km of beach, 3km, but symbolically it's a big deal, and the coalition that comes together, exists thereafter, protecting that beach in the long-term.

Over time, they've learned that surfers are particularly supportive of this work, and they have come to rely on the surfing community as critical partners in each project. In the coming years, Save the Waves Coalition is planning on training and upskilling dozens of champions from beach communities around the world to accelerate the work.

The water's edge is an exciting place for sports. Beautiful views, fresh air, awesome playing opportunities on land and on the water. Already, however, coastal erosion patterns are quickening and climate change is likely to accelerate that pattern in future, with more severe storms ready to strip away sand, and sea-level rise creeping slowly upward. I'm not saying it's time to move off the coast, but I am saying it's time for every coastal sports organization to start making contingency plans.

CHAPTER SIX

COME HELL OR HIGH WATER

In 2019, athletes at the Rugby World Cup in Japan waded through knee-high water to reach the pitch after Typhoon Hagibis tore through the region and dropped 240mm (9.5in) of water over Tokyo. It was the wettest storm on record in Japan and caused $10 billion (£8 billion) in damages, making it Japan's second most costly storm. This caught the attention of scientists Dr. Friederike Otto at Imperial College London and Dr. Sihan Li at the University of Oxford, who decided to calculate the extent to which Typhoon Hagibis could be attributed to climate change. Using weather observations from stations operated across the country by the Japan Meteorological Agency, the two scientists compared the likelihood of the storm occurring in today's climate, which has just over 1 degree Celsius of warming over pre-industrial baselines, with the likelihood of the same storm occurring in a hypothetical world with no warming. The results were stark: human-caused climate change increased the likelihood of extreme rainfall by 67 per cent in Typhoon Hagibis, reflecting the capacity of a warmer atmosphere to hold more moisture and thus drop more rain. Flooding from the extra rainfall accounted for more than 40 per cent of the damage. Climate change is costly.

Outside Japan, the typhoon was not big news... except that the Rugby World Cup was taking place. Instead of showing up on the front pages of newspapers and headlining the six o'clock news around the world, the typhoon got some of its most significant international coverage on the sports pages. As the death toll climbed above 100, the sports news focused instead on the wrath of the global rugby community, unhappy about a string of canceled matches in the early weeks of the tournament – none of which affected the ability of the impacted teams to move forward in the competition. Fans were reimbursed for their tickets to the canceled games, but travel expenses were lost costs. Sitting in hotel rooms in cities across Japan, thousands of unhappy fans who had traveled long distances to attend the tournament redirected their wrath to World Rugby.

Admittedly, World Rugby did put their flagship event in a country known for typhoons, in the middle of typhoon season. Contingency plans were in place, but they were weak: for the England–France match, for instance, the contingency plan was to relocate to a different stadium 14 miles away, which would change basically nothing in the context of extreme rainfall and flash flooding. To onlookers, the fans' response was unfair. Rugby is not nearly as important as the life-and-death impacts happening across the country. As Steve Busfield wrote in *Forbes* magazine, World Rugby was "not unsurprisingly, acting as though it is just a game and not worth dying for. Having thousands of supporters wandering around outside in a tropical storm is not behaviour to be encouraged." I tend to agree.

In the rugby world, the most engaged nations are all under increasing climate threats. For a six-week tournament slated to happen in October and November, there are few places that escape the wrath of storms, wildfires, droughts, and floods. In the past few years, parts of Australia have flooded each spring (which is October/November in the Southern Hemisphere). New Zealand's atmospheric science body, NIWA (National Institute of Water and Atmospheric Research), announced that 2022 was New Zealand's hottest year on record, and its eighth most "unusually wet" year. The start of 2023 was not much better, with Cyclones Gabrielle and Hale whipping the island in back-to-back weeks. In South Africa, the problem is drought – ongoing in some regions for more than five years. In the UK, depending on the year, it's drought or flooding. In the United States, the fall brings wildfires in the west and hurricanes in the south and east. Some of the smaller rugby-loving nations – Fiji, Papua New Guinea, Tonga – are too small to host a tournament of this size. Even if they had the facilities, it's unlikely a tournament on any of these islands would evade a storm between September and early November. In other words, there weren't many other good choices for a World Cup host country. Contingency plans for hosting in Japan just needed to be more robust.

Too much water

When I was collecting data for my doctoral dissertation, I interviewed sports directors across North America. The most common response I got to "What's your biggest environmental concern?" was "too much water". Flooding is tricky because there just isn't much you can do to fix it once the water lands on your field or in your building. You just have to… wait. The water will seep into the soil, fall into stormwater drains, or run off into the roadways and sewer systems nearby. But if the city's stormwater system is at capacity, your water has nowhere to go. Keep waiting.

If water sits on a basketball gym floor or a football pitch, it will wreak havoc on the surface. In Louisiana after Hurricane Ida, flooding in school and university gyms caused floors to become undulated, the hardwood shape-shifting into curved patterns resembling shallow waves. On a visit to the University of New Orleans athletics department in 2022, six months after Ida, I was shown to a gym which had just been renovated. Still, under the bleachers, evidence of the flooding remained. Paint peeled along the bottom of the wall, and the new floorboards lifted off the ground, forming what looked like bubbles in the flooring. On the fields outside, the storm pulled down fencing, destroyed goal posts and soccer nets, and waterlogged the grass. The whole sports multiplex was unplayable for weeks, and teams were sent to live, train, and play their home games in Memphis, Lafayette, and

Birmingham. To their credit, the athletics department was ready with a well-articulated emergency management plan, which ensured nobody was harmed, disruptions were minimal, and the rebuild happened relatively quickly.

In the UK, severe flooding caused by Storm Desmond in 2015 submerged Carlisle United FC's Brunton Park stadium under 3 meters (10 feet) of water. The damage was extensive, and the club had to play their next two home games at other nearby stadiums while the water drained. In the span of eight days, the staff re-laid the grass on the pitch, re-set the lower concourse where furniture had been smashed by the water, and brought in a set of generators for power because the electrical system was down. Besides, water and electricity don't mix.

Relocation or postponement are the preferred options during a flood, as these avoid additional disruptions to league play, tournament schedules, and other teams' plans. In November 2019, heavy rain and flooding affected many lower-league clubs, including Forest Green Rovers. Their League Two match against Plymouth Argyle was postponed due to a waterlogged pitch.

Sadly, in most cases, like the Rugby World Cup during Typhoon Hagibis, relocations aren't an option and tight scheduling makes cancellation inevitable. This solution ensures nobody gets hurt, and minimizes fan travel in unsafe conditions, but financially it's a painful decision. Losing broadcast revenues and issuing ticket refunds can be expensive and is not always covered under the "force majeure" clause of insurance schemes. Fortunately, where good insurance exists, cancellations can be done without major financial costs.

In 2023, the Formula One Grand Prix in Imola, Italy was canceled just two days before the racing was set to begin, owing to flooding. Many race staff, reporters, and fans had already travelled to Italy, but with flooded access roads to the race circuit, it was impossible to move people safely to and around the site. Thanks to the force majeure clause on their insurance policy, the race promoters were not forced to pay out their hosting fee, which would have been roughly $20 million. However, teams and media outlets had to foot the bill to move all their people home quickly, and arrange for their equipment (think of several large trucks per team) to leave Imola safely.

Beyond the costs of rearranging logistics in the case of floods, the cost of repairing storm and flood damages can be intimidating. In the UK, the cost of restoring the soil and replanting a single football pitch can cost up to £1.5 million, depending on the grasses or turf material being used. Taking up a damaged crushed-rock parking area and relaying new gravel will run to £170 ($215) per square meter/yard, and re-flooring a gym can cost upwards of £30,000 ($38,000), assuming the wood is available. There's been a global timber shortage in the last few years.

Through small grants of up to £15,000 ($19,000), Sport England has supported recovery efforts for impacted clubs. Eligible expenses include decontamination work, cleaning work, minor electrical works to restore power, or skip hire to remove sediment and trash. It's not usually enough to cover all the costs, but it will take care of some upfront expenses until the insurance kicks in. The England and Wales Cricket Board operates a similar small grant program for its sport. In Germany, after the mass flooding events that struck the west of the country in July 2021 and caused more than €100 million ($110 million, or £86 million) in damages to grassroots sport infrastructure, the German Olympic Sports Confederation approved an initial €100,000 ($110,000, or £86,000) in relief funds for recovery efforts. These quick funding solutions are a great start, but future flood damage may not be as fixable with stop-gap solutions. And no matter how you cut it, €100,000 is not going to resolve €100 million in damages.

Research led by Dr. David Goldblatt used Climate Central flood maps to determine how many existing football stadiums are at risk of flooding in a warmer future. By his count, 23 of the 92 professional football clubs in the UK are low-lying, coastal, or riverside, and could see partial or total annual flooding at their stadiums by 2050. Among them, Liverpool, Everton, and

Southampton are coastal clubs and could see flooding from sea-level rise. Manchester United's Old Trafford and Manchester City's Etihad stadiums are on the banks of the river Irwell and river Medlock respectively, which could put them at risk when the rivers overflow. Chelsea's iconic Stamford Bridge is near the river Thames, which could see flooding. West Ham United, playing in the relatively new London Stadium next to the river Lea, may also face flooding challenges. That's just one sport in one country. Twenty-five per cent of football venues will be potentially unplayable, which could be catastrophic for the leagues: you can't plan for a full season with that much disruption.

Dr. Russell Seymour, my colleague at Loughborough University London and Chief Executive of the British Association of Sustainability for Sport, suggested these flood projections are not particularly surprising. Historically, city planners in the UK placed sports grounds on flood plains next to rivers, as they were the largest plots of land available, supported healthy turf, and couldn't necessarily support heavy infrastructure because of softer soils. This did mean, however, that flooding was a possibility – even an expected reality. This was before sports grew more sophisticated in their expectations for turf conditions, amenities on site, and associated infrastructure like parking facilities, and way before professionalization turned sports into big business in the mid-twentieth century.

Elsewhere, similar fates await venues in major US markets like New York City, Miami, Chicago, and the San Francisco Bay Area. MetLife Stadium, home to the New York Jets and New York Giants, will be in a floodplain by 2050, facing annual flooding, along with City Field (New York Mets) and the Billie Jean King National Tennis Center (home of the US Open) in Queens. Miami could be partially submerged by high tides in 2050, with American Airlines Arena and LoanDepot Park wiped off the map. Cities on coastlines are already flooding.

When you think of where cities are located, traditionally, it's next to sources of water. Oceanfront, lakefront, and riverside locations made for good spots to settle in a time before water pipelines were conceived, and when land transport was best done on horseback. The biggest cities in every part of the world are on water. Roughly 40 per cent of the world's population lives in coastal areas, and a 2011 global study by Finnish and Dutch scholars published in *PLOS One* showed 50 per cent of the world's population lives within 3 kilometers (1.8 miles) of a surface freshwater source. There are of course some exceptions, particularly those cities that grew out of the railroad industry or from existing smaller villages that did not need international links: Atlanta, Mexico City, Johannesburg, Vienna, Bratislava. But the vast majority of humanity is on the water's edge, and those people living and playing on the same level as the water, with little elevation, risk experiencing flooding.

Looking around our cities today, it's hard to fathom them under water. I struggle to picture a football field looking like a giant swimming pool (of dirty water, but I'll get into that in the next section). Images of flooded schools, hospitals, and homes are something I associate with catastrophic events like Hurricanes Katrina, Harvey, or Maria. Grappling with this projection of a new normal is not easy.

When David Goldblatt published his report that included future flood projections in sports grounds globally, it fell on deaf ears in the sports sector. "It's unbelievable," he told me, "I emailed the supporters trusts at the football clubs most affected, and said 'do you want to talk about this?' Silence. I think there is a lot of denial. It will take something catastrophic for them to pay attention, or else a very big, very loud study that shoves the reality in their faces and makes it too important to ignore."

That catastrophic outcome may be closer than we'd like to believe. Some places may be unliveable in future, let alone unplayable. The most recent IPCC Report predicts that we will reach between half a meter and a meter (20–40in) of sea-level rise by 2100. What's scary is that the report writers specifically noted they cannot rule out a scenario of 2 meters (6ft) of sea-level rise in that same time frame, and 5 meters (16ft) by 2150. These massive surges would come from ice melt.

So far, glacier melt has accounted for only 21 per cent of sea-level rise since 1993, with Greenland's melting ice sheet contributing 15 per cent and the Antarctic ice sheet adding 8 per cent. History has shown us that it's possible for these stats to jump quickly. Ice sheet instability occurred repeatedly during ice-age cycles of the past, and could happen again. It's tricky to say exactly how fast the ice will melt and whether we might hit a tipping point beyond which chunks of glacier will break off and melt at an accelerated rate. Essentially, this would be a point of no return. There's not much we can do to prevent ice from melting. As scientist Stefan Rahmstorf has warned: we have enough ice on Earth to raise sea levels by 65 meters (213ft).

Sometimes, though, the flooding is slow and steady. At the Oakland Coliseum in California's Bay Area, home of the A's baseball team, if the stadium is flooding – and it is in a perpetual state of near-flood – the best-case scenario is to not have raw sewage leaking onto the field. More than once, the water pressure has grown so much that sewage pipes have burst, flooding the dugouts and clubhouses with human waste (imagine the stench!).

Dave Kaval, President of the A's, admitted the site is not ideally located. It sits on a depressed marshland, so it was always prone to potential flooding, but sea-level rise is making that worse. The Coliseum's playing surface sits a full 22 feet (6.5m) below sea level, and is almost completely surrounded by water. The banks of the Damon Slough and its tributaries run on three sides of the facility and are often at capacity during high-tide events under normal conditions. Add rain, and you've got flooding.

The A's are not just facing flooding from the outside. They're also facing flooding from the inside of the facility. Sea levels have risen 8.76 centimeters (3.5in) in the Bay since 1990, and the extra water has crept underground, pushing freshwater closer to the surface. This has put pressure on drainage systems, nearly reaching capacity even on a sunny day. With soaked soils underneath the baseball diamond, there's nowhere for extra water to drain naturally, so it pools on the field. In generations past, a set of expensive drainage pumps were installed under the stadium to capture and evacuate excess water. Staff used to turn them on a few times a year, when rain would pool on the field. In the last few years, Kaval admitted, "things are worse. The pumps are going 24/7 under the stadium. If we turn them off, the site would almost definitely flood. There's just too much water." Perennial floods do have a silver lining for franchise owners looking to justify a move. It's hard to argue that the current stadium is fit for occupancy, and inland locations are looking pretty good right now. As it so happens, the A's are moving to Las Vegas in 2024. Not just because of flooding – their lease was up at the Coliseum and a drawn-out process to secure building permissions for a new stadium were ultimately unsuccessful – but still, they're moving from a perpetually flooded site to the desert.

Even water sports are compromised by too much water (oh, the irony!). It's dangerous to play in high tides or on flooded rivers; the water is simply too powerful. In recent years, the Tacen Whitewater Course in Ljubljana, Slovenia, has twice been flooded beyond the point of safe paddling. Both times, the site was due to host an important competition. In 2017, just weeks before the venue was due to host the Canoe Slalom European Championships, more than 50 cubic meters (65 cubic yards) of equipment and material was washed away by the floods, including bits of the start gates and some of the rocks and other material that had been strategically placed in the water to influence the flow of the river for a higher quality competition. After the flood, it took days for the Sava River to run clear enough for the organizers to see what was in the water and how the flow would look for the competition, eroding the already limited time they had to rebuild the course.

The following year, the venue flooded once again, this time during the fourth event in the 2018 World Cup Series. More than 200 athletes from 30 countries had traveled to participate in the slalom race, and for two days, they did. But after strong competitions on Friday and Saturday, the semi-finals and finals were canceled because the water levels were too high and the river could not be safely navigated. Following competition guidelines, the results from the heats on Friday became the final results for the World Cup.

For other water sport athletes, floods have cut off their training seasons. In Ottawa, Canada, high water levels on the Ottawa River forced the local rowing club to cancel the whole high school rowing season in 2019, preventing the area's athletes from training for national championships. River levels reached the edge of the boat house, making it impossible to even install docks. The lost opportunities paled in comparison to the damages experienced by local residents whose homes were flooded, but to an excited group of 16- and 17-year-olds, it was devastating to lose their sport. Coach Cliff Brimmell told local news that the team had experienced something similar before: in 2017, floods in April meant the team couldn't get on the water until six days before the national championships. "We did make it just by the skin of our teeth." In 2020, the pandemic shut them down again. For the most recent crop of high school rowers in the region, it hasn't been an easy five-year stretch.

What's in the water?

But for all the impacts of flooding on land and on the water, the situation in the water is arguably even more dire. The water isn't just getting higher, it's getting dirtier.

"Every surfer knows not to go into the water after rain. You wait 72 hours, at least." Jake Howard is a surf journalist who has covered the sport for more than two decades for *Surfer Magazine*, *The Surfer's Journal*, *Red Bulletin* and ESPN. In California, where Howard lives, they can go months without any real rain. As rain or flood waters move over land, they pull chemicals off lawns, oils off the roads, and debris into the water.

Water quality is a growing concern in the world of water sports. The oceans have become the world's dumping grounds for waste of all sorts. Despite being a relatively new product, invented in 1907, plastic became ubiquitous through the twentieth century. Convenience and hygiene were two of the main reasons behind plastic's meteoric rise. The United Nations estimates that 400 million tonnes of plastic waste are produced every year, with 9-14 million tonnes entering aquatic ecosystems. Plastic makes up 80 per cent of all marine debris and water samples have found traces of plastic in every corner of the world. Including our bodies.

Cigarette butts, food wrappers, plastic bottles, bottle caps, grocery bags, and plastic straws are the worst culprits. It's impossible to measure exactly how much plastic is currently in our waterways, but some water sport athletes are helping scientists with data collection.

In 2017 and 2018, British sailor Dee Caffari skippered the *Turn the Tide on Plastic* boat in the 2017–18 edition of The Ocean Race. With a team of ten and a 65-foot boat, Caffari sailed around the world in nine legs, coming sixth in the seven-boat competition, but the trip was about more than just racing. Dee's campaign, backed by the Mirpuri Foundation and the Ocean Family Foundation, promoted the United Nations' Clean Seas campaign, which aimed to raise awareness of ocean plastics.

I interviewed Dee a year after the race finished, and she had vivid memories of seeing plastics just about everywhere along the route.

A lot of my team were doing this round-the-world sailing trip for the very first time, and they were commenting on how much they were seeing. So I realized that in my ten years of sailing around the world, six laps later, things are only getting worse. You're seeing balloons, crates, buckets, fishing nets, bottles, food packaging, and you're seeing it as you travel through the ocean. And on the stopovers we'd do beach cleans to get the locals engaged, and there you're seeing cigarette butts, you're seeing straws, and lots of little bits.

It's very easy to spot the macroplastics. But you can't see microplastics, which are tiny degraded bits of plastic that are in suspension in the water. So Dee agreed to participate in a science experiment as part of The Ocean Race's Science Programme. At regular intervals along the course, 96 microplastic samples were taken from the side of the boat by filtering seawater through three stainless-steel filters. In 93 of the 96 samples, they found microplastic particles, ranging from 0 to 349 particles per square meter of water.

The highest concentrations of microplastics were found in the West Tropical North Pacific Ocean (the South China Sea and West Philippine Sea), with 243 to 349 particles per square meter. Lower concentrations were found in the middle of the Atlantic and the Southern Ocean – far from the coastlines, but still, there was plastic. German researchers who analyzed the samples concluded in their summary study that "The results, with concentrations in the range of a few tens of particles per m^3 in offshore waters versus more than one hundred particles per m^3 in areas close to the continents, reveal that there is a general tendency for the microplastic concentration to be higher close to the continents."

Rivers, too, have their fair share of plastic. Lizzie Carr is a British environmentalist who made headlines in 2016 when she paddled 400 miles up the length of England's waterways on a paddle board, documenting the plastic waste she found along the way. It took her 22 days to paddle from Godalming in Surrey to Kendal in Cumbria. "I had to do something, and that was the only thing I knew how to do," she told Oceanographic Magazine. "I don't come from a science background. I know how to paddle board and I can definitely photograph and pick up litter. So that was the contribution that I could make."

I met Lizzie at COP26 in Glasgow, where we shared a stage for a panel on athlete activism. Her story was inspiring, morally driven, and one of the unique examples of recreational athletes who take up the mantle of environmental advocacy. In the eight years since she began this advocacy work, she's paddled across the English Channel, down the length of the Hudson River and launched the Planet Patrol app, which allows users to log the pollution they find in bodies of water near them. Through a partnership with the University of Nottingham, data collected on the app is analyzed and reports are being written on the types and distribution of plastic in British waterways.

In 2022, Planet Patrol started a citizen science water quality-testing project to address the fact that only 14 per cent of English rivers have "good" ecological status, down 13 per cent since 2010, and not a single English river has "good" chemical status per government standards. Despite these numbers, there is limited government funding going into water quality improvements and the country is off-track to hit its goal of having all rivers in good ecological status by 2027. So, short of funding and real commitment from the government, Lizzie and her team are building an army of paddlers to make small gains on the water quality agenda.

Surfers Against Sewage is also fully engaged on this issue, raising the alarm on the remarkably high amount of sewage being dumped into British rivers and the ocean each year, a problem that is contributing to the low water quality reports nationwide. In 2021, water companies discharged raw sewage into rivers 372,533 times for a combined total of more than 2.7 million hours.

Hugo Tagholm, Executive Director at Oceana UK and former CEO of Surfers Against Sewage, considers that British rivers have been treated like open sewers since the privatization of the water industry in the UK back in the 1980s. "These water companies had to remove sewage from the coastlines, so where did they turn? The rivers. But where do the rivers flow? To the coastline."

The 1991 Water Industry Act in the UK allows for sewage to be discharged into rivers only in extreme weather circumstances. The logic here is that if the sewers are overflowing, it's better to release that pressure somewhere downstream than to let sewage back up into people's homes and in the streets. So a concession was made that a small amount can go into the rivers, in extreme circumstances. But as Tagholm puts it, "The mentality in the water industry has been that anything that's not slightly overcast and 15 degrees is considered extreme. So if it's suddenly a bit sunny, they consider it drought, if it's suddenly a bit rainy, they consider it a flood. As a result the industry is really only set up to cope with the most bland weather situations." Consequently, the river-dispensing option is used far too often.

"They've got away with blue murder, quite literally," Tagholm lamented. Sewage pollution can cause algal blooms that starve rivers of oxygen and kill fish. Blooms can also impact mammals and birds that depend on the river for their food. When water isn't clean, it can also be unhealthy for water sport athletes. The pathogens in untreated sewage can threaten people's health, causing gastroenteritis or other forms of infection. Algae blooms can cause their own slew of health problems for people on the water: coughing, sneezing, itchy throats, and a variety of rashes.

Canadian Olympic kayaker Adam van Koeverden, now a Member of Parliament in Canada, remembers water quality issues impacting his training schedule on several occasions throughout his paddling career. "It was a huge problem, and it was always on my radar," he told a room of climate activists during a meeting on Parliament Hill in Ottawa, Canada, in April 2022. But one memory in particular stands out, because sport was both the villain and the victim.

In 2006, van Koeverden was in Florida for a Team Canada training camp on the Banana River, when the team started noticing something weird was going on. The first sign came from the athletes' GPS watches, which recorded faster times than usual. Way faster. Van Koeverden broke his 10km distance record by four minutes. Then the physical symptoms set in. Half a dozen athletes on the team fell ill with bronchitis, others had rashes, some had both.

Phone calls to the local water authorities and some internet searches provided more clarity. Chemical runoff from nearby golf courses had produced a chemical imbalance in the water that created the perfect conditions for a massive algae bloom to emerge. The algae created an emulsification on the water's surface, like a soapy water effect, which reduced the surface tension of the water and made it possible for the boats to cut through it with little resistance. The algae also gave the athletes rashes and bronchitis, disrupting the rest of the training camp.

While athletes playing on the rivers, lakes, and oceans are aware of these issues, Tagholm says "the majority of people don't act on it." Dr. Rebecca Olive, Senior Research Fellow in the Social and Global Studies Centre at RMIT University (Royal Melbourne Institute of Technology, Australia), agrees: "surfers and swimmers may be aware of the issue, but they don't always address it."

Olive has spent the last decade studying open-water swimmers and surfers to understand their experiences, so she can speak to their interests and help them change their minds. Her body of research explains that what we eat, what we wear, and our consumption choices all end up reflected back in the water, mixing and flowing with the other contents of the ocean: the bacteria, the plastic, the fish, and the animals that may sting or bite if we get too close. And though it may not be comfortable to think about, it's important to consider the ways that we're not just benefiting from water sports and ocean environments, we're also harming them. Our presence has an impact: the microfibers of our wetsuits, the oils on our bodies, our spit and bodily fluids, all play a role. "That's not to say that we shouldn't go into the water, far from it," she says, "but we do need to be more considerate about how we share these spaces. It can't just be about the benefits the water brings to us, the conversation also has to be about how we can benefit the water."

In recent years, the conversation is starting, and not just by environmental activist groups. Surf and boating brands like Notox are coming out with products like the Ecoboard, made from organic, natural, non-toxic, and recycled materials. Sportswear giant Patagonia has released wetsuits made from Yulux, a neoprene-like material that's derived from natural rubber harvested from rubber trees that are Forest Stewardship Council-certified by the Rainforest Alliance. And a variety of reef-safe sunscreens and body lotions have been released by the beauty industry. But it'll take far more upstream work to reduce the sewage, plastic, chemicals, and other toxic products entering our waterways globally. That's a fight that will likely happen outside of sport at government levels. What sports participants can do is take care of their own water, like Planet Patrol, and organize politically – and loudly – around legislation that will reduce upstream spills, like Surfers Against Sewage.

CHAPTER SEVEN

DUSTBOWL

By the cruel logic of the hydrology cycle, a climate-changed future will see more flooding, but it will also lead to more drought. For sport, both are a nightmare.

In 2016, as drought swept across India, the Bombay High Court ordered that all Indian Premier League (IPL) cricket matches scheduled to be played in the state of Maharashtra after April 30 be moved out of the state. The decision impacted 13 games, including the final, which was due to be played in Mumbai on May 29. Stating that the court "cannot act as a mute spectator and ignore the plight of people", the judgment aimed to conserve water and prevent the use of large amounts of water for maintaining the cricket grounds while residents of the city went without showers.

The logistics of relocating so many games led to a mess of last-minute changes and travel plans. Rescheduling efforts were rushed and stressful for those involved. The IPL has sufficient budget to accommodate such changes, being the second-richest sports league in the world, with upwards of $10 billion (£8 billion) in annual revenues. Cricket outside the IPL had a harder time with the droughts. The Board of Control for Cricket in India (BCCI) voluntarily canceled two Ranji Trophy matches that were scheduled to be played in Maharashtra, and reports of water scarcity and drought impacting cricket pitches elsewhere in the country also stopped play at youth levels.

Maharashtra experienced its worst drought in a century in 2016, leading to restrictions on water use across a range of industries. Despite being an economic driver for Pune, Nagpur, and Mumbai, professional cricket is not an essential service, so it was not exempt. It's hard to justify bright green pitches when it comes at a cost of several million litres of water per season, and while residents are being told to take shorter showers and hospitals are being forced to cancel non-urgent operations.

IPL leaders argued that it could use treated waste water to maintain the pitches, but this option was dismissed by the courts. Two teams – the Mumbai and Pune franchises – offered to donate money to drought-relief efforts, but this was not enough to dissuade the courts from stopping play either. The decision was frustrating for the league, which attracts some of the world's top players. But water cannot be bought, especially not in the volumes the cricket teams need it, when a whole country is experiencing successive years of drought and crop failure. To illustrate how bad the water shortage had become, a train carrying half a million litres of drinking water was sent to the worst affected area of the state.

It is hard to explain the sense of despair that comes with water shortages. In an elemental sense, humans need water to survive. Academic research has found links between water shortages and mental illness. A 2015 meta-analysis led by Dr. Holly Vins at Emory University in Georgia, US found evidence that drought's adverse economic effects can compound stress and social isolation. The study explains that when drought impacts the economics of an area, business opportunities, and expenses for households, that's when things get very challenging. This is particularly true in rural areas, where economies are more likely to be tied to agriculture. In 2015, more than 3,000 farmers in Maharashtra died of suicide, the highest number this century, with many of the deaths attributed to stress over crop failures. But there could be a similar link for sports. The research hasn't been done – but if those most reliant on the land are most impacted by drought, many outdoor sports would fall in that category.

Climate change has not been kind to water supply. As hydrologist Professor Taikan Oki wrote, "Water is the delivery mechanism of climate change's impacts to society." For much of Asia, including the Indian subcontinent, which gets its water from the Himalayas, Western Ghats, Eastern Ghats, Aravali Range, and Vindhya Mountains, the melting glaciers are drying up the water supply. More than 800 million people are at least partly dependent on meltwater from these glaciers. Even in a world where temperature increases are limited to 1.5 degrees Celsius – which seems unlikely at this point; it's going to get hotter than that – around one-third of the ice stored in these glaciers would be lost by the end of the century.

Drought in Cape Town

India is not the only place at risk of severe drought. In 2018 – just a year after India's multi-year drought ended – the Indian cricket team was visiting Cape Town to face South Africa, and they were told to take two-minute showers. Cape Town was in a historic drought of its own. The city gets nearly all its water from dams that rely on rain, and from 2015 to 2018 the rains didn't come – or at least, not as much as they usually do.

It's worth noting that Cape Town was not wasting water. In 2007, the City of Cape Town implemented a comprehensive water management program aimed at minimizing water waste that won the city the prestigious C40 Cities Award for Adaptation Implementation in 2015. The program included a public awareness-raising campaign on water use efficiency, a plumber training program, and a range of technical improvements to the municipal water system including pressure management schemes, leak detection, and pipe replacement programs. When the drought hit in 2015, they were ready.

Initial efforts to reduce water use were effective. People took shorter showers, watered their lawns less – and then not at all. Households saved the dishwater from one meal to wash up after the next. Then, as the drought stretched on, restrictions increased. At the beginning of January 2018, new water restrictions were implemented that limited residents to 87 litres per person per day. One month later, the limit was dropped to 50 litres per person per day – for everything: cooking, cleaning, bathing, laundry. Under these restrictions, there was nothing left for sports fields.

At the beginning of 2018, when water levels hovered between 14 and 29 per cent of total dam capacity, the city started planning for Day Zero: when the water would hit 13.5 per cent of dam capacity – a critical low. At that stage, municipal water would effectively be shut off, and residents would have to queue at 149 collection points for their daily rations of 25 litres per person. Fortunately, Day Zero never came, as heavy rains in June 2018 replenished the city's water supply. But the drought got bad enough and lasted long enough to shut down sports across the city.

The 2017/18 Western Province Cricket Association (WPCA) premier league, first, second, third, and reserve divisions all canceled their seasons owing to the drought. Sports fields were left in complete disrepair after weeks of not watering. At the time of cancellation, the WPCA services manager, Clinton du Preez, was quoted in local news outlets saying the organization was "forced into this position" because of the deteriorating conditions, and that they'd "seen the quality of cricket drop over the course of the season due to the poor state of our fields and pitches".

Several other field sports were also impacted. Across Cape Town, football venues closed to conserve water, and the five Premier Soccer League teams based in the city played out of the Athlone Stadium, putting a lot of pressure on the pitch. If the drought had lasted any longer, it would have been unplayable. When the Hamilton Rugby Ground hosted the Cape Town Rugby Tens tournament in February of 2018, water trucks delivered drinking water, chemical toilets were on site, there were no showers, and no water misters or splash pools for cooling.

At the Cape Town Cycle Tour in March 2018, drinking stations were supplied by water trucks brought in from outside the city, and locally produced desalinated water was used for cleaning. Event organizers estimated that visitors to Cape Town would use between 1.5 and 1.7 million litres of water during their stay, so they purchased 2 million litres of spring water from sources outside the city, which were reintroduced back into the municipal system.

The Two Oceans Marathon held at the end of March 2018 relied on purified spring water drawn from a borehole in Newlands, an affluent suburb of Cape Town. Boreholes are narrow, deep holes drilled into the ground to access water, oil, gas, or other subsurface resources. They provide a source of clean water for drinking, irrigation, and other uses, and are commonly used in areas where surface water sources, such as rivers and lakes, are limited or contaminated. As boreholes access water that is outside the municipal supply, they are not subject to municipal restrictions and are a popular option for sports fields and golf courses, where they're available. At the Two Oceans Marathon, the water was distributed to runners in little sachets. No showers were offered at the finish line.

So some sports carried on, to some extent, but the Cape Town drought impacted the scheduling and delivery massively. At lower levels, sports just stopped. School sports and youth sports clubs don't have the money to truck in water or pay for desalinated water to keep their pitches healthy. And there is good evidence to expect similar, possibly more severe, and more frequent droughts in South African cities in future. If global warming passes the 2 degrees Celsius average global temperature increase, a threefold increase in the risk of severe drought is projected across the country. And of course, playing on hard surfaces is dangerous and can increase the risk of injury.

Worldwide responses to drought

Dr. Dave Rennie is the head of medical performance at Bristol City Football Club in the UK, where he oversees all injury prevention, recovery, and safe training practices. He earned his

PhD from Liverpool John Moores University, where he studied how pitch hardness impacts injury risk within football. "Dried-out pitches, like water-logged pitches, can be dangerous for the athletes," he told me over a call. "People will say it's a good leveler when the pitch is in poor condition, either too wet or too dry or frozen, etcetera, but really that's a stand-in for saying it's unhealthy to play on."

Think of it like walking on a beach. If a pitch is hard, it promotes game speed. You have to put less energy into the surface to get the same energy back and propel movement. But it's also inflexible to fall on, and tougher on the joints, so we see stress injuries, or more severe injuries from collisions and falls. Meanwhile, soft pitches are like soft sand. They are less damaging to the joints because there's more absorption of energy, but they're also exhausting to run on.

Maintaining the right amount of moisture in the soil is critical to performance and injury prevention in sport. Rennie continued, "A huge amount of water is needed to maintain healthy turf. If there's too little water, and the pitch gets dry, we have to change the way we practice to reduce injury risk."

In Australia, the amount of freshwater available in rivers and lakes could decrease by 10 per cent as a result of a 1.5 degrees Celsius global temperature rise. To cope with the shrinking water supply, the biggest cricketing attraction in the Southern Hemisphere – Melbourne Cricket Ground (MCG) – has a clever way to conserve fresh water: it doesn't use any. Instead, MCG relies on recycled water from its own water treatment plant located underground just next to the facility. The Yarra Park water recycling plant treats local sewer water to the highest standard for recycled water. The water may not be good enough to drink, but it's certainly good enough to maintain the pitch and clean the facility.

In many ways, sports have pioneered the use of water recycling for grounds management. In Scottsdale, Arizona, a region so water-scarce it is classified as a desert, a similar water recycling plant is used to supply water to golf courses. In the 1980s, Scottsdale's city council was looking for ways to reduce water use in the city, having experienced some drought years. One solution was to ban the use of drinking water by golf courses, which can use 2–3 million gallons of water a night. In response, 12 of the city's golf courses banded together to co-fund a water treatment facility and a 13.5-mile-long (22km) pipeline to transport that water from the plant to the fairways. The whole system took only a handful of years to set up, and by the mid-1990s, every course in the city was using recycled water. So when Grayhawk Golf Course uses 2 million gallons (or roughly 9 million litres) of water in a night, know that it's not taking water from the drinking water supply. It's taking water from the sewer system.

In Rio de Janeiro, Maracanã Stadium was outfitted with a state-of-the-art rainwater harvesting system before hosting the FIFA World Cup in 2014 and later the Olympic Games in 2016. While most people were paying attention to the array of solar panels added to the roof, the less conspicuous sustainability upgrades on the building involved installing 18 rainwater filters to the edges of the 50,000sq m (60,000 sq yd) roof, which captured and cleaned the water running off the roof and redirected it to a large harvesting tank. Between the 18 filters, the system could capture up to 288 litres per second in heavy rain, supplying enough water to maintain the pitch and fill the 292 toilets in the facility. Overall, this system reportedly reduced the facility's reliance on externally supplied water by 40 per cent.

Another effective response to drought is reducing water use. Watering fields and pitches can be reduced in the off-season, while the team is playing away, or when there has been rain. Watering less can make a difference in water consumption, but it also makes a difference in soil moisture. Grasses aren't the real problem – it's dry soil that turns the surface hard and can reduce an athlete's ability to get traction under their boots. Minute changes in soil moisture can make the difference between a player getting traction and making the play, or losing their footing and missing an important pass or tackle.

Most sports fields have moved to a night-time watering schedule, to avoid evaporation during the warmer daytime hours. Hand watering is also a way that facility managers can control exactly how much water is being used on each part of a site. It's possible that some areas need more water than others, for instance, if the site has different grasses – which is common on golf courses, for example, which can have several grass varieties planted on site. Or perhaps there are parts of the field with different use profiles, such as the pitch, infield, and outfield in cricket. Each of these areas requires different amounts of water. It's also possible that some areas of a field are shaded, while other parts are exposed to consistent sun during daylight hours. Wind can also play a role.

Computerized systems have emerged to offer efficient watering solutions to groundskeepers at scale. Much like the sprinkler systems some people have at their homes, the computerized watering systems at sports grounds can ensure water is distributed evenly and at the most optimal time of day; early in the morning, just before sunrise. Additional technologies can be added, like moisture sensors in the soil, which trigger watering when they get too dry. These sophisticated systems can also be set to different moisture levels at different parts of the ground. For instance, you might water less on the sidelines and in the end zones of an American football stadium, and more in the main part of the ground, where athletes are likely to do most of their running, turning, and tackling.

Lord's Cricket Ground in London has pioneered water management solutions in cricket that stave off drought. In 2022 when the city was a dustbowl, Lord's was lush. Like other major cricket grounds, including Melbourne Cricket Ground and Centurion Park in South Africa, Lord's uses a computerized sprinkler system to distribute water efficiently and to ensure no excess water is wasted. It's called the Rippleffect system, and it monitors water use not only on the fields, but in each building on site, and has been instrumental in helping the facility managers identify leaks and opportunities for water reduction. Since installing the system, Lord's discovered that actually, most of the water use is happening in the buildings, not on the field. So additional steps were taken to ensure that toilets and urinals have water reduction controls such as flush sensors.

While most solutions to stave off drought make it possible to retain natural grasses, some turn to artificial turf, which can save money and resources over the long term. Traditional artificial turf requires very little water, except to alleviate some of the heat retention of the surface. Most of the research on injuries has been done on these surfaces. Newer iterations of artificial turf, called hybrid turf, seed natural grass into a fabric of artificial turf to ensure the natural grass is fully supported. These hybrid turfs do require water because they are partially grass. However, player associations across a range of field sports have been vocal about their preference for fully natural fields, including the NFL Players Association.

Citing NFL data from 2012 to 2018, J.C. Tretter, president of the NFL Players Association, wrote that the rate of non-contact injuries is significantly higher on artificial turf than on natural grass. The NFL disputes this, maintaining there is no clear difference, particularly on newer hybrid turf products. Of course, there were no major water shortages between 2012 and 2018 that reduced NFL stadiums' access to water, so we don't know what the stats would look like on parched fields.

As I write this, the US Southwest is in a historic water shortage that is causing inter-state conflict between California and Arizona, which are grappling in a legal battle over access to limited water supplies. The last two decades have been the driest in 1,200 years, and while rain at the beginning of 2023 offered some reprieve, the soils were so dry that they couldn't absorb all the water and runoff led to massive floods. The region's capacity to adapt to these dramatically different water realities is slim.

Mitigating the crisis

At the center of the water crisis, the Colorado River, perhaps best known for carving the Grand Canyon over millions of years, is running dry following decades of infrastructural development and dams, a growing population, the expansion of large-scale industrial farming in the region, and climate change. The river starts in the snow-capped Rocky Mountains – which are fast losing snow – and snakes its way across the western US for more than 2,000 kilometers (1,240 miles) and over the border into Mexico, serving as the main water source for more than 40 million people. It also supports a recreation industry of boating, fishing, hiking, and other outdoor activities on its two reservoirs – Lake Powell and Lake Mead – worth an estimated $26 billion (£20 billion). Both reservoirs are shrinking. According to research by the Scripps Institution of Oceanography, the Colorado River's average annual flow could decline by 30 per cent by 2050. Unless water use is dramatically reduced over the next two decades, Lake Mead has a 50 per cent chance of declining to "dead pool" by 2036. At that level, water supply to millions of people in California and Arizona and to the region's farmlands could be cut off. It would also stop hydroelectric production at the Hoover dam, which provides electricity to Las Vegas and Southern Nevada, and to Southern California.

And so, as California Governor Gavin Newsom moves forward with plans to create new storage for stormwater, and Arizona Governor Katie Hobbs goes about modernizing the groundwater supply and closing loopholes that have allowed water to be poached by private firms without paying, policymakers in every watershed are looking for ways to reduce water use. There are a range of policy options available to governments to reduce demand on the dwindling water supply. These include implementing a system for water drawing permits, revisiting water pricing, and providing incentives for private firms to co-fund water capture and reuse systems.

In most cases, water restrictions and permitting systems are put in place when water resources are low. However, it's worth noting that there are differences between residential and commercial water restrictions. If a business relies on water for the upkeep of their facilities, they are usually exempt from some of the more extreme restrictions. Sports venues and golf courses will typically fall into that category. For instance, in California, the State Water Resources Control Board adopted new restrictions in the summer of 2022 to stave off water shortages. The restrictions require urban water suppliers to create a water contingency plan for how they would adapt to shortages of 10 to 20 per cent. Water suppliers in California are also required to report annually on water use and demand, and to implement a rule or ordinance limiting landscape irrigation with potable water to no more than two days per week and prohibiting irrigation between 10am and 6pm. However, sports fields and recreation areas were excluded from the policy.

Even in the UK, a nation known for rainy weather, drought is creeping into the picture in recent years. In 2022, a prolonged period of extreme heat and summer drought turned sports pitches brown. From my office window in Queen Elizabeth Olympic Park, I could see brown fields in the park itself, and dull tones of beige and brown at Hackney Marshes just up the river on my way to work. I watched one afternoon as children playing rugby and football came off the pitches with bruises and scratches from falling on rock-solid grounds. July 2022 had been the driest month in England since 1935. The use of hosepipes and sprinklers was banned due to water shortages, leading sports fields to go without water. Breaking a hosepipe ban could land a groundskeeper a £1,000 fine – a steep price when sport at the youth levels is already underfunded. But as the total stock of water in England's reservoirs dropped to 65 per cent of normal capacity in midsummer – the lowest level for that point in the calendar year since 1995 – something had to take the hit.

Sports fields and golf courses are not the biggest users of water; that would be the hydro villains over in Big Agriculture. Irrigation for agriculture accounts for 70 per cent of water use worldwide according to the Organization for Economic Co-ordination and Development (OECD). In contrast, golf – the thirstiest of sports – uses only tiny fractions of the overall water supply, even in the driest states: 1 per cent of California's water and 2 per cent of Arizona's supply is used on golf courses. And in most cases, water conservation practices are in place to reduce the amount of potable water being used for watering fields and greens with a view to dropping those percentages further in the coming decade. But with increasing pressures on total water supply, it's likely more water restrictions will come in future years, as we saw in Cape Town in 2018. And with more demands on water from boreholes, rainwater, and recycling plants, sports may need to start sharing those supplies as well. So, while sports have come a long way in innovating around water use, we have a long way still to go if we want to keep our fields safe for play in the future.

———————

There is something sports can do that most other sectors can't: raising awareness of water shortages and galvanizing the public around water conservation strategies. Manchester City FC and their sponsor Xylem, a water company based in Switzerland, have figured this out. In a recent short film shot at the Etihad Stadium, Manchester City players Phil Foden, Raheem Sterling, and Sergio Aguero help depict a future world without water.

The three-minute film centers on a young fan on her way to be a player escort at a 2020 match. She watches the morning news while eating her breakfast as presenters discuss climate change and warn of drought risks. She then brushes her teeth with the water running, walks by a leaking pipe on her way to the stadium, and arrives at the edge of a bright green pitch with roaring crowds in the stands. Then the film jumps a generation. The year is 2045 and the girl is grown up with a son of her own. The same sequence repeats itself as the woman takes her son to serve as a player escort for a City match, but with no water in sight. In this depiction of the future, street vendors are selling water bottles for £20 ($26), restaurants are boarded up because of water shortages, and the pitch is parched and beige. In somber tones, the announcers state that due to the sustained drought, this could be the last match City plays.

The film is grim but based in reality. The Environmental Protection Agency has predicted that by 2045, due to ageing infrastructure, increasing demand, and climate change, parts of England could run out of water. Internationally, the outlook is worse – England being a relatively wet nation compared to many in deserts or arid regions. According to UN Water, already some 1.2 billion people live in areas of physical water scarcity, while another 1.6 billion face "economic water shortage". By 2025, almost half of the world will be living in conditions of water stress. Pulling on the heartstrings, the Manchester City campaign ends with a call to action.

"If we act now, we can save our future and the game we love."

CHAPTER EIGHT

SHELTER OF LAST RESORT

When Hurricane Katrina made landfall at 6am on August 25, 2005, its impacts were swift and devastating. Originally pinned as a violent Category 5 storm, the hurricane was downgraded to a Category 3 storm by the time it reached New Orleans – but the damage was already done. When the storm moved over the Gulf of Mexico before sunrise, it caused a storm surge that pushed forward over land, engulfing the lower 9th Ward of New Orleans under 8ft (2.5m) of water by breakfast. The pressure from the surge was enough to break the levees around the city, sending residents racing for shelter. About 90 per cent of the city was able to evacuate. But for those who couldn't leave, the Superdome, the Louisiana Convention Center, and the Riverwalk Shopping District became the last line of defense.

The city wasn't ready for this storm. Most of New Orleans sits below sea level, with its levees designed to withstand some water, but not this much, and certainly not this fast. If the levees had held, the city would have been saved. The storm was also not supposed to hit New Orleans – it was originally heading for Florida. So, in the days leading up to August 25, the city's leaders knew a storm was coming but weren't prepared for anything of this magnitude.

For the roughly 20,000 people who sought shelter in the Superdome, conditions were abysmal: power and sanitation in the building was limited, and people slept in every nook and cranny – on the field, on the steps, in the grandstand, along the

halls of the concourse. It was crowded and desperate. Glenn
Menard, General Manager of the Superdome at the time, stayed
in the facility with his wife Jill to help oversee the emergency
response. In a conversation published in *The Advocate*, Menard
explained that the facility staff worked with the National Guard
and the New Orleans Police Department to deliver a shelter
program, which included providing food and water. This proved
especially challenging when, instead of receiving the 300,000
ready-to-eat meals they'd been promised, only 41,000 meals – all
served cold – were delivered to the dome. When the roof tore off
partway through the storm, which caused water to pour into the
facility and added to the pressure from water flooding outside,
things got worse. According to one radio reporter, Dave Cohen,
who was in the Superdome during the storm, it became a "place
of mass suffering".

By August 30, five days into the emergency, flooding rose up
to 15ft (4.5m) high in some parts of the city and the stadium was
still in a state of desolation. Louisiana Governor Kathleen Blanco
ordered the evacuation of the site, transporting the remaining
evacuees to the Houston Astrodome on 68 buses. The Federal
Emergency Management Agency (FEMA) had reportedly
promised more than 400 buses, so the evacuation took five days.
The Superdome was not completely evacuated until September
4, 2005.

Over in Houston, preparations for the mass shelter at the
Astrodome were far more organized and efficient, buoyed by extra
time, functioning infrastructure (roadways, electrical systems, and
phone lines), and the fact that the Astrodome was publicly owned.
As John Spong wrote in *Texas Monthly*:

The best measure you'll get of how poor the response to
Katrina was in New Orleans – no matter which level of
government you choose to blame – is how quickly the
Astrodome was readied. At three o'clock Wednesday morning,
August 31, with New Orleans still filling with water, Harris

County Judge Robert Eckels was awakened by a phone call from the coordinator of Governor Rick Perry's division of emergency management. (The Dome is under Harris County jurisdiction, making Eckels, the county's chief executive, the man to call.) By six o'clock Eckels was at the home of Reliant Park's general manager, and the two of them were working with Mayor Bill White and a host of others to make the relief effort happen. The Astrodome had seen only occasional use in the past six years, and the first step was to get it up and going. That morning the air-conditioning and plumbing were upgraded. While that was happening, the Red Cross shipped in tens of thousands of cots, blankets, and "comfort kits", little bags containing toiletries ... When the first buses arrived at ten that night, the Dome was ready.

These two stories illustrate the importance and utility of advance planning and coordinated response efforts for using sports facilities as emergency response sites. Sports facilities are certainly big enough, and have most of the right amenities to quickly convert into shelters, but this doesn't automatically make them good or sufficient. Significant coordination is required between the facility management, local authorities, government agencies, voluntary organizations active in disaster (VOADs), and emergency services.

Be prepared

Twelve years after Hurricane Katrina flattened the Gulf Coast, and sports facilities were used as shelters and emergency response sites, a similar approach was adopted to respond to Hurricanes Irma and Maria in the Caribbean – with better results.

Since the US and Canadian media didn't cover Hurricanes Irma and Maria nearly as fully as those that hit the continental United States, less is known about the experience of the hundreds of people who lived inside the Coliseo Roberto Clemente for nearly three months during and after the storms. To understand the scope of what happened there, I spoke with Carmen Yulín Cruz, former Mayor of San Juan.

> Puerto Rico has been in a state of perpetual emergency for many years. San Juan is a frequent site of flooding, and those in poorer neighborhoods at low-lying areas are most susceptible to damages. I activated the largest sports complex in Puerto Rico, the Coliseo, which has basketball and gymnastics and other sports, and the baseball ballpark next door, and another smaller indoor coliseum. When Hurricane Irma was coming our way on September 6th, I moved my executive and emergency services departments inside the coliseum because it's a place that doesn't get flooded, so we could keep generators and food and supplies there. In a way, we knew how to prepare and respond, because our city is a frequent site of flooding. When Hurricane Maria hit two weeks later, with the city still without power or most functioning infrastructure, the coliseum became not only the largest shelter in Puerto Rico, it was also the place from which I managed the command center for all emergency response.

During Hurricane Maria and in the immediate aftermath, 685 people sheltered at the Coliseo Roberto Clemente along with 200 city employees. Down the street at a smaller sports facility, an additional 400 people sheltered with 50 city employees. A third facility served as a storage site for generators and powerlines that would provide energy when the power went out.

These facilities weren't perfect shelters. For instance, the showers were insufficient, since in most stadiums the shower facilities in locker rooms are designed to serve a group of (usually) men who know each other well and travel together. It's no big deal for them to shower in the same room, but in an emergency shelter, when you have men, women, and children who don't necessarily know each other and who are in a state of distress with few personal belongings and limited access to hygiene, these showers fall short. People in emergency shelters need privacy, which is hard to come by in large, shared spaces. To address this, Cruz and her team had a circus tent with makeshift shower stalls erected next to the building a few days after the storm. It wasn't perfect, but it was private.

In terms of supplies, Cruz's emergency management department ordered two months' worth of food and emergency medical supplies so that shelters would be covered for the duration of the storm. What they didn't anticipate was needing to share those resources with the rest of San Juan. Sixty-five elderly homes became reliant on the city's emergency management team for food and water when their supply chains fell apart. While people typically remember the destruction caused by Hurricane Maria, the issue was that the island had been hit by Hurricane Irma two weeks before and resources were already depleted.

In emergency shelters, morale is fragile. Stress and anxiety levels are high, and in the absence of internet and working phone lines, people in the shelter can't communicate with their loved ones or find out if their home is still standing. The common denominators are grief and uncertainty. To address this, the San Juan Sports and Recreation Department worked alongside the Culture Department to offer daily sports programming for children and adults who were sheltering inside the Coliseo. Using the space for its intended purpose, and enjoying the opportunity to play on professional-caliber basketball courts, brought some much-needed positivity to the community that had formed inside the shelter. After everyone had left, Coliseo Roberto Clemente became the largest distribution center of aid in all of Puerto Rico and remained operational in this capacity for several months.

Reflecting on her time living in the stadium shelter, Cruz offered this:

> It was an experience I will never forget. It gave a new meaning to what those facilities are for. There was a lot of pain and

struggle and despair in the place. So my job was not just to feel that despair, but to compartmentalize that and to provide hope and the semblance that things were on a schedule.

I always thought that sport and culture were a pillar of social development, but Irma and Maria gave that a whole new meaning. When I came into my position as Mayor, there were a dozen sports clubs, and I tripled that. There were 39 when I left. I always perceived sports facilities and clubs as the heart of the community, and the community was ready for that: these programs were turned over to the community. The people became in charge of creating their own programs, and that really worked. To see the Coliseo used in a different way was sad in many ways, but inspiring in others. We realized that in order for us to move ahead and come out of the tragedy, we needed to be the hope, to embody hope. And those sports facilities provided space for that.

The other thing about sports facilities is that people feel comfortable in them. People come together over sports, they feel comfortable there. So when push comes to shove, I'm going to feel more comfortable when I go to a place I know and recognize and that I associate with good memories. This can also happen in smaller sports facilities, local gyms, basketball facilities, gymnastics. I think we could do better by the taxpayers if we make sure that we are able to transform the building from its primary usage to a different use in a time of crisis.

In addition to the Coliseo and the other major sports facilities offered as shelters, the smaller community sports facilities around the city became the central arteries of the resource distribution network. Although it didn't receive press coverage, Cruz's team set up a municipal resource distribution system during the recovery period using outdoor basketball courts around the city. Each week, at designated times, water and food would be delivered to neighborhood residents at the local basketball courts, where doctors and nurses would also be on site to provide medical care. As Cruz put it, the basketball courts were selected as distribution sites because "In any given community, everybody knows where the local parks and basketball courts are. So in the midst of the chaos, it was easy to communicate the location of supplies and ensure that people knew where to go." This made sense, especially when you consider that people's phones and digital maps wouldn't have worked while the power was down.

After Hurricanes Irma and Maria, Cruz bought a 2500kW generator the size of a small bus that could provide electricity to the whole Coliseo in the case of future storms and shelter situations. But she still isn't satisfied, because that's only one building that's ready for another disaster. Most people can't fathom worst-case scenarios, but in Puerto Rico, as in New Orleans, they lived it. Cruz and her fellow Puerto Ricans know the generator is a good step but not enough.

When I went to visit the Coliseo four years after the storm, I met with a facility manager named Juan who took me on a tour of the building. The first thing that struck me was how dark it is inside: stadiums are designed to draw everyone's eye to the court or the field, not the seats or the concourse. Of course, there are lights, but it feels like a proper bunker in there without the noise and the scoreboards and the buzz of a live game – dark walls, a dark ceiling, no source of natural light in the main bowl. The next thing that becomes clear is just how little privacy is afforded to residents in super-shelters like stadiums (or any shelter, but it feels *especially* open in a super-shelter). Juan demonstrated how the stadium stands could be collapsed to make more room on the floor of the facility for camp beds, and showed me just how far apart each one was set up from the next. It was tight; had I been staying there, I could've touched the next bed while lying on my own.

As he was showing me around, Juan shifted his tone, almost as if he were annoyed. He explained that people should never have been housed here. "Yes, it worked, but it wasn't pretty.

Everything was not used right." He explained that it took several months to get the facility back up and running as a sports venue after the sheltered people moved out, because the facility was so degraded by their presence. When I asked what he might suggest as a better response, he said, "We need to use our smaller sports spaces for housing, and the bigger facilities like this one for resource distribution and FEMA headquarters and the clinic. Or just better planning so we don't have to improvise as much."

Earlier in this book, I discussed the damaging 2020 wildfire season. In the western US in 2020, more Americans spent nights in emergency shelters such as community centers and local gyms than in any other year of the previous decade. During the 2020 federal election, sports facilities were turned into voting sites. In some cases during 2021, they were also used as COVID vaccination sites. Despite these "wins" for stadium use in emergency situations, it is important for the media to tell these stories gently, and to avoid implying that these facilities are being used as efficiently as possible for these reactionary purposes. They're not (although voting sites and vaccination clinics are the exceptions to the rule, because they benefited from advance planning). The responses to natural disasters are simply patchwork methods, and better ones are possible. Under the pressure of collapsing infrastructure and the climbing death rates so common when hurricanes, fires, and other natural disasters strike, a proactive system is needed to ready sports facilities for emergency uses.

How do we make sure our facilities have solar panels and reliable access to power in the case of grid shortages and power outages? What about shower facilities, or the provision of medical support and clinical materials in shelters? How can we rewrite the script for emergency responses to ensure communities at risk will know where help will be distributed ahead of time?

The damage associated with hurricanes is worsening.* According to the National Oceanic and Atmospheric Administration, "of the 310 billion-dollar weather disasters between 1980 and 2021, tropical cyclones [or hurricanes] have caused the most damage: over $1.1 trillion [£870 billion] total, with an average cost of almost $20.5 billion [£16 billion] per event." The Association of British Insurers estimates that by the 2080s, climate change could cause a 75 per cent increase in costs of damage in a hurricane season in the southern US, a 65 per cent increase in costs of insured damage in a hurricane season in Japan, and a 5 per cent increase in wind-related insured losses from extreme European storms. Both 2011 and 2017 saw record-setting hurricane seasons with 16 events across the southern US, and in 2020 the record was beaten with 22 events.

Even the simple practice of naming storms got complicated in 2020 because of how many there were. The World Meteorological Organization (WMO) uses peoples' names to identify storms: for Atlantic hurricanes, an international committee puts together six lists of 21 names, which are then rotated so that each list is repeated every seventh year. Male and female names alternate on each year's list, in alphabetical order, skipping the letters Q, U, X, Y, and Z because there aren't enough names that begin with those letters. In 2020, the whole list was exhausted, and we entered the Greek alphabet for naming storms: alpha, beta, gamma, delta, epsilon, zeta, eta, theta, and iota. It became so confusing for the public (consider how similar "zeta", "eta", and "theta" sound, and they all formed within days of each other) that the WMO

---

*A *tropical cyclone* is a generic term used by meteorologists to describe a rotating, organized system of clouds and thunderstorms that originates over tropical or subtropical waters and has closed, low-level circulation. In the North Atlantic, central North Pacific, and eastern North Pacific, the term *hurricane* is used. The same type of disturbance in the Northwest Pacific is called a *typhoon* (US National Ocean Service).

has decided it will not use the Greek alphabet again. Instead, a 21-name backup list has been developed for use in years when the initial list doesn't cover the full season. Let's hope we don't see that backup list. But in all likelihood, we will.

While there is no conclusive evidence that the duration of the season or the number of cyclones will increase (it should hold relatively stable around 80 per year, as it has since 1985), the proportion of Category 4 and 5 hurricanes could increase at a rate of about 25–30 per cent per degree Celsius of global warming. Storms may also slow, unleashing more rain and wind on the areas they hit, and shift poleward, expanding the geographical range of the regions impacted by storms. The destruction and disruption associated with hurricanes could be harsher in the future; all available resources should be organized proactively and deployed efficiently to preserve human life and community wellbeing through future disasters.

Where does emergency help come from?

In the United States, FEMA is the main source of disaster relief aid. However, their responses to crises are mostly reactionary. While some grant money is available to support proactive disaster prevention and mitigation efforts, these funds pale in comparison to the funds spent on relief and responses after the fact. A major overhaul is needed.

In 2010, the International Association for Venue Managers (IAVM) and the American Red Cross co-published a "Mega-Shelter Planning Guide". The guide emphasizes the importance of advance planning and collaboration among stakeholder groups, including the sports organizations who own and run the facilities. However, according to Harold Hansen, one of the Guide's editors and the former Director of Life Safety and Security for the IAVM, "most people in city governments, like in the sports and entertainment sector, assume the worst-case scenario will never happen to them, so planning a response is not a priority."

There are a number of conflating factors that further complicate collaborative efforts, including ownership of the building (private or public) and the reimbursement of funds by FEMA and other state and federal sources. It's important to note that governments don't like to transfer money between federal and state levels, so if a city- or state-owned sports facility opens as a shelter, the federal government may not reimburse the local authorities for that effort, assuming it should be covered by city or state budgets. In contrast, when a private facility opens as a shelter, they'll almost always be reimbursed completely for expenses related to operating the shelter. This creates tension among sports managers who want to ensure they aren't compromising the future financial viability of their facility by opening their doors in times of need. They must decide whether to help now and risk hurting financially later, and often this decision is made with limited time and under pressure.

According to Dr Samantha Montano, emergency management expert and co-founder of the Center for Climate Adaptation Research, the permissions needed to open shelters can be easier to attain among publicly owned facilities, because city officials (or state officials, in some cases) can make that call without consulting private partners. If taxpayers fund the facility's construction and some of its operations, then city officials have a reasonable justification for using the facility for emergency use. In North America, however, many sports facilities are owned by the teams, so the complicated stakeholder matrix is unavoidable.

It's important to note here that the use of sports facilities as mega-shelters is not a proposition that compromises the current core purpose of these facilities: businesses can host events and make money on game days, in conventions, and whatever else is happening in the facility. Mega-shelters (and their smaller counterparts in school gyms and local community centers) will only activate when disaster strikes. So the relevant question with regard to potential lost income is: how many sports events are being held when the city shuts down due to an emergency?

Spending money and time now on preventative planning for emergency facility use will save money and lives in the long run. This method could even open avenues for stadiums to be fortified against storms or other hazards through public grant money, assuming FEMA will eventually shift their funding priorities and offer cities more resilience-building funds. But again, it's important to bear in mind that it's not a case of whether disasters will come; it's a question of when and how extensive they'll be.

The added cost of designing a stadium to meet mega-shelter standards should, in theory, be relatively low (with the possible exception of a roof structure). Modern engineering standards mean most stadiums already have the structural capability to safeguard residents during a public emergency. A few things are unclear and require further clarification and codification: the conditions that would trigger an emergency takeover, the legal liability of housing residents and ensuring their safety, and the cost of operating the mega-shelter (and event cancellations). Previous attempts to define such provisions (for example, requiring sustainable design certification and community benefits agreements) have failed to gain traction. Still, cities are investing more and more in efforts to reduce their climate vulnerability. As the public cost of these mitigation and adaptation strategies is similarly expected to grow, a proactive stadium mega-shelter declaration would not only complement a city's broader climate (and economic) policy, but also strengthen the stadium's status as a public good.

With American cities likely to "continue to (1) build stadiums and (2) grapple with the effects of climate change", my colleagues Tim Kellison (Florida State University) and Nick Watanabe (University of South Carolina) have repeatedly argued that benefits to the public should be maximized. With public funds being used to fund stadium infrastructure, these stadiums should not be reserved for events and accessible only to those who can afford the price of a ticket. At least some portion of a stadium subsidy could be justified if city leaders assured the public that these spaces would double as a mega-shelter or a resource distribution site during emergencies, or perhaps a voting site during elections, or a mass vaccination clinic if the need arises.

Since Hurricane Katrina, New Orleans has completely overhauled its emergency response plan for hurricanes and other extreme weather events. Following the devastation, US Congress authorized and funded the Hurricane and Storm Damage Risk Reduction System, which built and strengthened 133 miles of perimeter around New Orleans and surrounding areas to protect it from storm surges and to pump out excess water. The 911 emergency system also underwent a sweeping overhaul that consolidated medical, police and fire calls under the New Orleans Office of Homeland Security and Emergency Preparedness. From a structural perspective, there were several important measures taken to improve resilience across New Orleans. But among the most iconic and morale-boosting efforts was the Superdome rebuild.

The rebirth of New Orleans and the Superdome's revival are almost synonymous. Repairing and re-opening the facility a year after Katrina sent a message to everybody that they can rebuild, too. But it wasn't a straightforward path. In the first trips back to the dome, senior members of staff were dressed in hazmat suits, walking around dogs and cats that had been left behind by evacuees, or were strays that had found shelter in the stadium. Structural experts estimated the rebuild would cost more than $200 million, assuming mould could be removed. It was a grim starting point, but NFL commissioner Paul Tagliabue saw it as an opportunity to reinvent the space and rejuvenate the city's downtown. So he encouraged stadium managers to think of it as a fresh start, and it worked.

After a full season playing away from New Orleans in 2005/06, the Superdome hosted the Saints' first home game of the 2006/07 season to a packed house. The city came together around the football team in a building that had been called a "monument to suffering" just a year earlier. The Saints, in their successful 2006/07 season, galvanized city pride to witness an iconic punt block by Scott Gleason that would later be commemorated with a statue in front of the building, marking the beginning of the Saints' and the City of New Orleans' new, post-Katrina chapter.

The Superdome isn't used as an emergency shelter anymore. In fact, after the gruesome situation in the city during Katrina, officials have decided to cut all in-city shelter services and to focus instead on evacuations.

Enter the Smoothie King Arena. Home to the Pelicans in the NBA, the facility has become central to evacuation plans as the primary staging area. The city has grown in size in recent years, so it was determined that the former staging area, Union Passenger Terminal, was too small. With the Smoothie King Arena, the city can quadruple the number of people it can process. An estimated 35,000 to 40,000 people will use this service, roughly 10 per cent of the city's population. To access the evacuation service, residents can make their way directly to the Smoothie King Arena or go to one of 17 "evacuspots" spread across the city, each marked by 14-foot (4m) steel structures outside the door, where buses will collect them and take them to the Smoothie King Arena to be registered in a central evacuation system and matched to a bus leaving town for a shelter elsewhere. Because the Smoothie King Arena is state-owned, the city doesn't have to pay for its use in emergencies. It just has to cover the costs of any damages incurred by its use as an evacuation staging site.

The rebuild in New Orleans has been anything but equitable. There were several major failings during the storm and in the rebuild that have left poorer, majority Black areas of town in dire straits compared to their richer, white neighbours. But the new recovery plan, which provides free evacuations in the three days ahead of the storm, is an improvement for everybody and targets those most in need. It's far from perfect, but it's progress.

In Puerto Rico, the rebuild after Hurricanes Irma and Maria was less straightforward. As of January 2020, two and a half years into the recovery, only a measly 4 per cent of the obligated FEMA funds for sport and recreation facilities had been disbursed. All across the island, communities were still without recreation and sports opportunities. Where they did exist, they were piecemeal and makeshift.

On top of being frustrating for residents who didn't have access to adequate facilities for years on end, the delayed rebuild presented a bigger problem when, in January 2020, a series of earthquakes sent thousands seeking shelter. The municipal stadiums in Yauco, Sabana Grande and Guayanilla served as shelters for people affected by the earthquakes, despite not being in the best condition themselves. This was an unwelcome reminder that sports facilities can only provide shelter from the storm if they're well maintained between the storms.

CHAPTER NINE

LET IT SNOW

Mornings in a ski town have a certain routine swell of excitement to them. In homes and hotels, the lights flicker on around 7am – before sunrise, with just enough time to eat some breakfast, grab your gear, and head to the lifts. In the 20 minutes before the lifts open, ski towns look like ant farms. Swarms of skiers converge on the sidewalks, walking slowly in uncomfortable boots, the swish of snow pants providing a background to the traffic. As the sun rises over the peaks, it creates a halo effect over the mountain top. There's nothing quite like it.

The culture of alpine snow sports goes back roughly 100 years. That might not sound like much next to the 400-plus-year history of golf or surfing, but the passion shared by skiers and snowboarders runs deep (pun intended). Sit in an Austrian bar during the winter season and at least one television will show the FIS (International Ski Federation) World Cup competitions, even if they're just repeats from three days ago. I could wax poetic about skiing for 200 pages of this book. It's easy to do – the mountains are magic. But with bad winters becoming more common, they're also endangered.

In July 2022, I spent a week on a writing retreat in the Alps, attempting to tap into the inspiration that started me on this journey. I got more than I bargained for. Tignes is a ski resort that stretches from a base at 1,400m (4,600ft) to a peak higher than 3,700m (12,000ft), topped by the Grande Motte glacier. This is one of Europe's finest resorts. It has hosted Paralympic events

(the Albertville Games in 1992) and Winter X Games, in addition
to countless FIS Ski World Cup events. It boasts more than 300
kilometers (about 200 miles) of pistes and some of the most "snow
sure" conditions in the Alps. It's also widely known as one of Europe's
premier summer skiing destinations. But as I walked through the
town, signs at the lift ticket office announced the glacier was closed
to skiers. A few weeks earlier, they'd offered some summer skiing,
but the conditions deteriorated fast due to high temperatures and
they had to shut down just a couple of weeks into the season.

The following winter, at the start of 2023, in the last months of
writing this book, I went back to the French Alps to ski with family.
The region had enjoyed amazing early-season snow in November
and early December, followed by rain over the holiday season and
warm temperatures that temporarily shut down half the resorts in
the Alps. By mid-January, the snow was back, but not enough: it
took herculean efforts by the snowmakers to produce the snow
needed to open all the runs. And then, between late January and
the end of February, they got no new snow at all. The conditions
were devastating – a mix of ice and slush, depending on where the
sun met the mountain. The valleys were green, the mountain tops
white, thanks to snowmaking. The trends are too obvious to ignore.

That week in the Alps, on what was meant to be a ski holiday
and a break from work, I spent my time moping around, fighting
off sadness. I had to remind myself not to comment on the lack of
snow in every interaction, lest I come across as a total downer. My
favorite sport, and my favorite place, and one of the main sites of
my research career as a sport ecologist is dying. This feels personal.

———

The impacts of climate change on snow sports in the Alps, in the
United States, and in Canada are well documented. The ski season
is getting shorter, and snow conditions are becoming unreliable.
In the US, for example, one study published in 2018 in the
journal *Geophysical Research Letters* offers evidence that average

annual snowpack has diminished by 41 per cent since the 1980s. In Europe, the same pattern exists. A study by the Institute for Snow and Avalanche Research estimated that Swiss ski resorts have lost over a month of snow coverage since 1970, with the season starting 12 days later and ending 26 days earlier. November skiing is gone completely; April skiing is not far behind.

There's a "100-day" rule of thumb in skiing. Owing to all the expenses associated with operating a ski area, including staffing, marketing, and exorbitant utility bills associated with running chair lifts and snow guns, a ski resort needs around 100 open days per season to be financially viable. We're getting dangerously close to that number on both sides of the Atlantic, especially at low-altitude resorts.

Natural snow isn't quite as plentiful as it once was. Snow data compiled by the Rutgers Global Snow Lab shows a steady and significant drop in Northern Hemisphere spring snow cover, from around 32 million square kilometers (12 million square miles) in 1967 to about 28.5 million square kilometers (11 million square miles) in 2015 – a 10 per cent decrease in just four decades. And while most mountain resorts have adopted snowmaking technologies, even the best solutions have limitations.

Currently, the best available estimates suggest that 95 per cent of ski resorts rely on snow guns for at least some of their snow. In extreme cases, typically at lower-altitude resorts, artificial snow can account for 80 to 90 per cent of the snowpack. Snowmaking is a good solution for when it's cold out but there isn't much snowfall. The challenge is this: you can produce all the snow you want, but if the temperatures and ground are too warm (typically, above about 5 degrees Celsius, 41 Fahrenheit), the snow won't survive once it leaves the snow gun. In other words, snowmaking is an excellent stopgap solution for low precipitation, but it won't solve high temperatures.

The potential damages associated with low-snow winters are significant. Conservatively, in 2016, the winter sports industry added $11.3 billion (£8.9 billion) to the American economy through lift tickets, hotel stays, travel, restaurant meals, and gear purchases, supporting nearly 200,000 jobs. Between 2001 and 2015, in the years when snowfall was low, the industry's economic contribution dropped by over $1 billion and about 10 per cent of jobs were lost. Those figures represent only the US. Other countries with major ski tourism markets, like Canada, Switzerland, Austria, France, and Italy, feel the economic hit of bad snow years too.

Training the professionals

The 100-day rule that keeps skiing viable as a tourism industry isn't enough to keep elite athletes in play. The typical World Cup circuit for skiing starts in early November and runs through early April. But there are summer training camps and pre-season training on top of that. So the actual needs of the competitive ski and snowboard calendar are closer to 200 days.

Phil Marquis competed in the 2014 Sochi Olympics and the 2018 PyeongChang Olympics for Team Canada's freestyle team. He's now a coach for the national freestyle ski program. We met working on Green Sports Day activations together in Canada. Marquis's concern with a shorter season is that if the season is shortened too much, we could see injuries rise and innovation fall. Despite improvements in technologies and training techniques off the snow, like trampolines, water ramps, and the use of air bags, he explains, "There are certainly concerns about the lack of on-snow preparation and training before a dense and heavy competition season. The physical requirements to compete at the highest level have to be incrementally achieved to sustain the demand and reduce injury risks."

Summer skiing is important to the training calendar. For Canada's mogul skiers,* the group Phil coaches and used to

---

*Mogul skiers take part in a discipline of freestyle skiing where riders race down a bumpy course, completing a series of jumps on complex terrain.

compete with, the goal is to stay on the snow until July. Thanks to high-elevation ski areas in the Rockies in places like Whistler, British Columbia and Mount Hood, Oregon, that's traditionally been possible. Then, in August and early September, the focus is on fitness and water ramping* so athletes are ready to return to snow in the fall, when they start training in late September until the season begins in November. Those on-snow opportunities have also been tough to plan between Europe, South America, and North America because of the unpredictable weather patterns in those regions over the past decade.

In 2022, the Canadian mogul team was planning to ski on the Hintertux glacier in Austria for their first pre-season training block, scheduled in September. Communication with the mountain's management team was great and a large quantity of snow had been conserved over the summer under giant tarps for the early fall training, supplying snow for the various ski and snowboard disciplines that were due to train there. But by August, it became clear that Europe's abnormally warm summer had caused everything to melt. Even the high-altitude glaciers were affected, losing massive quantities of year-round ice. Unfortunately, the mountain team had to dig into their covered snow supply to fill some newly formed crevasses in the glacier over the summer, and weeks before traveling, the call came through that the Canadians wouldn't be able to train there. All the travel plans had to be changed at the last minute, a nightmare the coaches and management didn't need.

Fortunately, Marquis explained, "We had tracked the weather and snow accumulation in South America, and relocated our training camp to Chile within a matter of days." Getting all

---

*In water ramping, skiers or snowboarders take a jump into a pool of water, which allows them to practice their jumps on more forgiving terrain and extend their training season into the spring and summer months when snow is hard to find.

the pieces into place so quickly was a stroke of luck. But it had to be done. Without those pre-season training opportunities, competition outcomes during the season would be compromised. Athletes won't push themselves to throw the biggest possible jump or the most complicated trick if they haven't had time to practice it enough on snow. If pre-season skiing starts to taper off, we could see some of the innovation in the sport dwindle – fewer new tricks, fewer impressive jumps, and possibly more injuries.

The other challenge is figuring out how to manage a choppy season once it's started. In the last few years, because of COVID and some bad snow seasons, up to half of each season's World Cup events have been moved, rescheduled, or canceled. Decisions to cancel tend to happen fairly last-minute, usually after travel plans have been made, sometimes even after the athletes have arrived at the resort, as organizers hold out for a turn in the weather. That kind of uncertainty can ruin an athlete's mental readiness and focus. It can also mean the total available prize money in a season is lower, as races typically don't get rescheduled when they are canceled. As for sponsorship, it can be hard to retain sponsor dollars when the opportunities for exposure are diminished by canceled events.

In 2023, responding to the poor conditions and repeated cancellations on the World Cup circuit, nearly 200 professional and retired skiers wrote an open letter to the International Ski Federation (FIS), demanding action on climate change. This wasn't the first time a group of elite and professional athletes had sent a letter to their sports federation: rugby players had done the same in 2021. So there was some proof that this type of action worked. The letter included the following:

As FIS athletes we are already experiencing the effects of climate change in our everyday lives and our profession. More and more often competitions have to be canceled due to extreme weather events or lack of snow. Pre-season training slopes are getting rarer and shorter every year, because glaciers are shrinking at a frightening pace. A heatwave in January brings the next lack of snow to Europe. Soon we won't be able to produce artificial snow at some classic World Cup slopes anymore as winter temperatures rise above zero in low altitude ski resorts more frequently. The public opinion about skiing is shifting towards unjustifiability. This will also bring the industry into trouble. Our sport is threatened existentially and urgently.

They followed up their statement with a request for action by the federation, specifically pointing to a need for a clear net-zero commitment, a sustainability strategy, and more transparency on this agenda at the federation level.

Other recommendations in the letter included shifting the season to start and end later, reorganizing the calendar to reduce transatlantic trips mid-season, adopting sustainable practices at events, committing to political advocacy on climate change, and offsetting any unavoidable emissions. It's a hefty list, but other sports federations such as FIFA and the International Biathlon Union have already made similar changes, so it's all possible.

Days after the athletes' letter was sent, FIS responded with a statement, disagreeing with the athletes' assertion that not enough was being done on the climate agenda. They pointed to their new president's career-long work in the climate space, and also to recent behind-the-scenes work to conduct a carbon accounting exercise and sign the Sports for Climate Action Framework, which will see them halve emissions by 2030 and achieve net zero by 2040.

Frankly, the FIS's response was not a good look. "We're already doing something about this," in a season with multiple weather-related cancellations and resort closures, is not good enough.

———

Even the Winter Olympics, the biggest show on snow, is on the line. At Sochi 2014, 85 per cent of the snow was artificial, because the area just doesn't get enough. PyeongChang 2018 was worse, with roughly 90 per cent artificial snow. Beijing hosted the Winter Olympics in 2022 with 100 per cent artificial snow. Not exactly ideal.

The athletes had mixed reviews of the situation. Some like the artificial snow because it's more consistent than natural snow, freezing at warmer temperatures and staying hard throughout the day. This means that if one athlete goes down the hill at 10am, and the last athlete goes down at 2pm, they're competing on the same surface. Others found the conditions unpleasant and risky.

But as I've shown in this chapter already, artificial snow might be the only option in future as fewer and fewer places have the natural conditions to host competitions without it. In fact, having an artificial snow system is now a requirement for hosting, according to the International Olympic Committee's guidelines, because they know it's necessary to ensure there's a Games at all.

Dr. Daniel Scott at the University of Waterloo and his colleagues have produced studies predicting that only one of the 21 cities that have previously hosted the Winter Olympics would be able to reliably provide fair and safe snow conditions for the ski and snowboard events by the end of this century. Sapporo, Japan will be the only option: a destination far from the majority of winter sport athletes, who are clustered in Europe and North America.

Fortunately, the study delivered good news, too. If the Paris Climate Agreement emission targets can be achieved, the number of climate-reliable host cities jumps to eight. In a statement, one of the study's authors, Dr. Robert Steiger of the University of Innsbruck, wrote "Climate change is altering the geography of the Winter Olympic Games and will, unfortunately, take away some host cities that are famous for winter sport ... Most host locations in Europe are projected to be marginal or not reliable as early as the 2050s, even in a low emission future."

Rising prices

Winter sports have a few other elephants in the room to address if they're to be saved in the context of a climate-changed world. One is the climbing cost of lift tickets. Glen Salzman has been skiing at Mont Tremblant, north of Montreal, since the late 1950s. His father worked as a ski boot maker. He remembers a time before chair lifts, when T-bars would pull you up the mountain, suspending kids mid-air because they weren't heavy enough to stay on the ground. The lifts were painfully slow. It could take up to 30 minutes to travel the same distance modern chairs cover in seven, so the staff would wrap skiers in wool ponchos to keep them warm on the way up the mountain.

The Mont Tremblant of Salzman's youth looked different than it does today. There was a single lodge at the top of the mountain – which is still there and now serves as the patrollers' hut – that served a short menu. Homemade pea soup and hot dogs, the same two offerings every day. The runs were narrow and not groomed so you had to pick your way down the mountain carefully. A day ticket for adults cost around Can$10 (US$7 or £5) in the 1960s, but you could get single-ride coupons for cheaper. Accounting for inflation, that works out to about Can$100 today (US$70 or £50), per the Bank of Canada's inflation calculator.

Fast forward 70 years and things look different. Since Intrawest bought the resort in 1991, the vibe in the town and the experience on the mountain has transformed. There are now several large chalets, a fast eight-seater gondola and four-seat high-speed chair lifts, a tourist village, sprawling parking lots, and year-round activity. The development has generally been good for the local tourism economy. But lift prices have gone up. Across the ski industry, prices have outpaced inflation. If you go to Tremblant for a day of skiing, it'll cost you Can$145 (US$107 or £85) per adult. A day on the mountain for a family of four – not including equipment rentals or lessons – will set you back about Can$500 (US$370 or £290) before tax. This is not exactly affordable for

Tremblant locals or day-tripping Montrealers, where annual median salaries sit around Can$60,000 (US$45,000 or £35,000).

The cost increases have a lot to do with rising operating costs, of course. Installing fast chair lifts and dozens of snow guns is not a simple feat, and that equipment needs regular maintenance by professional operators. The cost of energy and water has also gone up. But those explanations don't make the ticket any easier to afford for the average family.

Over time, resorts across North America have been taken over by a growing number of ski conglomerates. Companies like Vail, Alterra, Aspen, and Powdr have bought up resorts all over the world, creating unlikely partners of mountains on both coasts, sharing the operating costs and creating economies of scale in their purchasing. There are pros and cons to this arrangement. On the one hand, it's driven up the price of day tickets and eroded the local culture that makes each mountain town unique. For instance, resorts like Mont Tremblant in Quebec used to be smaller, more community-focused, and more French. But as Intrawest took over, and later Alterra, the resort started to look more and more like a copy-and-paste of Steamboat or Killington in Vermont.* But for those fortunate enough to be able to afford season passes, prices are good, and inter-mountain passes like the Ikon Pass, Epic Pass, Mountain Collective, and Indie Pass have increased tourism among those who can shell out hundreds of dollars at the top of the season for the privilege of traveling around and skiing in multiple resorts.

Lack of diversity

The other problem is that many – too many – skiers are white. This is obvious to anybody on the mountain. According to

---

*These resorts have very little character, instead offering visitors a soulless selection of chain restaurants and big-brand stores.

Snowsports Industries America, non-white skiers accounted for about one-third (31.3 per cent) of all skiers in the 2019/20 season, but only 27 per cent of those who skied more than three times in the season. Among them, only 3.3 per cent were Black. The "Unbearable Whiteness of Skiing", as University of Notre Dame Professor Annie Coleman called it in her 1996 article by the same title, was not a mistake, it was a design feature of ski resorts in the American West – with similar trends observed in Europe. As immigration to America expanded through the twentieth century, American skiing traditions were shaped by the experiences and preferences of European immigrants who brought their ski culture with them. Tracing the labor history of the ski sector, Coleman wrote that as the industry was being established in places like Colorado, Utah, and California, it centered on the ethnicities of white European immigrants and pushed people of color into back-of-house roles, serving, cleaning, and maintaining the facilities for the white recreationists.

All across the industry, the names of resorts, hotels, ski shops, and restaurants, shouted the industry's intentions to mimic a European experience. Coleman identifies countless examples: "Hotel St. Bernard, the Alpenhof, the Innsbruck Lodge, and the Edelweiss", restaurants serving European food on the menu (think schnitzel, goulash soup, Swiss fondue), and stores advertising Scandinavian fashions. The other central element of European whiteness in resorts were the ski instructors, hired internationally from Austria, Norway, and across the Alps, who served as the face of the sport. Still today, if you go to a mountain or check out the website for a ski resort, you'll see white faces on the mountain and in the marketing material.

Henri Rivers is the President of the National Brotherhood of Skiers, established nearly 50 years ago with a mission to identify, develop, and support athletes of color to compete at the highest levels of snow sport. The Brotherhood now has 52 clubs across the US, and international chapters in the UK and Canada. In an interview on my podcast in 2020, Rivers shared a message for ski resort managers: "Diversity, equity, inclusion, are very good things. It's not like you're opening the doors and doing people favors, this is about their survival." He went on to say, "To be divisive is not sustainable. To be racist is not sustainable. To be exclusive is not sustainable. And if we want to see the ski industry survive, we have to market to those communities that are underrepresented on the mountains and bring them in. There are millions of individuals that would love the opportunity to ski."

Local difficulties

For those who live and work at ski resorts, there are other fish to fry. Mental health in mountain towns has long been a problem: suicides rates are high. It's hard to explain why you're feeling down when everyone else is boasting about how amazing their day was. So mountain towns have a lot of self-medicating people with depressive symptoms. In the US, the ten states with the highest rates of suicide are the mountain states: Montana, Alaska, Wyoming, New Mexico, Idaho, Utah, South Dakota, West Virginia, Arkansas, and Colorado. Is that all to do with skiing? Of course not. But it can't be ignored.

Deirdre Ashley, Executive Director of Mental Health & Recovery Services of Jackson Hole in Wyoming, pointed to a range of issues that compound in the mountain states to make it hard to address mental health issues. Some of these challenges include a conservative culture that stigmatizes mental health and admissions of weakness, and the distance between mountain towns and the nearest major health centers, making it hard to seek more intensive care if that's needed. However, she also identified ski culture as attracting risk-takers and adrenaline-seekers. "In Jackson Hole, there's a culture of extreme sports," she told me over the phone, "and so, you know, there's also the added aspect of risk-taking behavior that is very much ingrained in that culture, and it's drawing a certain type of person who may be more at risk of developing mental health issues and self-medicating by taking those risks and seeking that high. When you add in the party culture, it's everything to the extreme, for sure."

Ski towns are not just expensive for visitors. Living there can be expensive, too. As Deirdre explained, "We have the highest income disparity in the country and the cost of living is very high. A lot of people will come here, and figure out that they might have had a vision of what it was going to be like, and then they're saddled with the reality of the cost of living. And in order to make it work, people are working two or three jobs and they're not doing the things they came out here to do. Or they're doing a very limited amount of it. And they're far from home, far from their support systems, and isolated." It's easy to see how that kind of gap between expectations and reality can lead some people to despair. The same trend can be found abroad: Queenstown, New Zealand – a picturesque ski town on the South Island – brought in a permanent psychiatrist after a sharp rise in suicide calls to the local police in 2018.

Alcoholism and drug use rates in ski towns are high compared to other towns of the same size and demographic composition. A 2017 Swedish study revealed that ski resort employees had a higher risk of drug and alcohol abuse than the general population. The risky use of alcohol was 82.9 per cent for seasonal snow workers compared to 58 per cent for non-seasonal (year-round full-time and part-time staff), and 8.3 per cent drug use in the seasonal sample compared to 2.8 per cent among the non-seasonal group. Of course, alcoholism and drug use are a risk factor for suicide: a 2020 study led by Dr. Frances Lynch at the Center for Health Research in Oregon found that those with alcohol, drug, and tobacco use disorders were at a 30.7 times increased risk of suicide.

If those stats alone aren't scary enough, things are getting worse with each season now being unpredictable in length. Early-season excitement is curbed when the snow doesn't come till Christmas, or later. For those who get their joy – and paycheck – on the snow as instructors, lift operators, coaches, or the hundreds of seasonal jobs in hospitality on and near the mountain, the absence of snow hits hard. Shifts get cut or shortened, and fewer people get hired overall. A new form of depression is now entering the already crowded arena of diagnoses among the skiing community: solastalgia.

Originally coined by Dr. Glenn Albrecht and his research team at the University of Newcastle in Australia, solastalgia means a new form of depression centered on grief and loss around the way things used to be, environmentally speaking. It's prevalent among farmers, for instance, and indigenous communities: those whose livelihoods and cultures are most closely tied to the land. That includes people who work in the snow sport sector.

Dwindling opportunities

For those too young to remember the ski days of old, climate anxiety and climate grief are the relevant terms to describe their justified sense of despair as they face down a future that may not sustain this valued cultural practice. While their parents managed to carve out careers in the mountains, entering at the lowest ranks as ski lift operators and catering staff, and moving up into supervisory and managerial roles, eventually reaching the upper echelons of mountain management and eking out a living in expensive resort towns, the younger set of mountain workers likely won't see those opportunities present themselves. Sometimes, in the worst cases, ski resorts just can't be maintained at all, with climate change piling onto a range of other constraints and ending resort operations altogether. In cases like these, the locals are left without work and live with a crumbling social infrastructure as other business start to shut down, too.

Jeremy Davis is a ski historian who runs the New England Lost Ski Areas Project, which he started in the 1990s. For nearly three decades, he has compiled data from newspaper archives, old brochures and advertising materials from now-closed resorts, postcards, magazine articles, and countless pictures and emails sent in to his website from more than 50,000 people. As he tells it, the heyday of ski areas in the north-east US was in the 25-year period following World War II, so roughly the mid-1940s to the end of the 1960s. The growing middle class was gaining access to cars, and the stable nine-to-five work week meant most working adults had weekends off, creating the possibility of day trips and weekend jaunts to the nearby mountains. Ski areas were popping up across the north-east US region, mostly small family-run operations. Some were even non-profit and volunteer-run: local Lions Club or Kiwanis Club members would set up small ski areas in their town to serve the local families. It was an exciting time. The growing ski business led to chair lifts being put in place, moving ski areas away from rope tows. By the 1960s snowmaking machines came onto the scene, ensuring good conditions on bad weather days.

In the 1970s, things changed. The gas crisis made a day trip to the mountains more expensive for the core segment of middle-class families that had been responsible for the boom in the 1950s and 1960s. Insurance rates for activities like skiing skyrocketed, driving up operating costs at every ski area. The Vietnam War broke out, pulling many men into a contested war effort. And competition rose from other destinations: because air travel became more affordable, skiers could fly to the Rockies or the Alps, or choose to spend their leisure dollars in other ways. Disney World and similar attractions drew families year-round, including March Break visitors, which ski resorts had previously relied on for a financial boost partway through the season. And then 1979 and 1980 hit, delivering consecutive seasons of poor snowfall. The 1980 Lake Placid Winter Olympics became the first to use artificial snow, and hauled in snow from nearby areas, but there just wasn't much natural snow to draw from. Many resorts in the region started losing snow days in the 1970s and while snowmaking propped them up for a while, it wasn't enough to keep them going when the temperatures started to rise, too. The operating costs linked to snowmaking started to outweigh the balance sheet. Not enough people were coming.

When these ski areas closed, the locals felt their loss. According to Jeremy:

a lot of these places were community gathering spots. You had people vacation there for years, some of them met their spouses there, their kids learned to ski there. They went to beer leagues and parties, and season kickoff days and youth competitions, or they worked at the mountain. And so these were much, much more than just a place to get some exercise in. They're places where you see your neighbors and make new friends and everybody has fun … and it's kind of tragic, in a way, that when you lose it, it's probably not coming back. And it hurts the local economy, too.

Creating snow

The technology standing between most ski resorts and permanent closure is snowmaking. Originally invented in a lab in Japan, perfected and finessed into machines in the US, and rolled out commercially in the 1950s and 1960s, snowmaking has grown to become arguably the most important technology in snow sports. It's hard to overstate the importance of snowmaking. In some places, it has extended the length of the season, ensuring spring skiing continues. In others, like the US Midwest, snowmaking is used to ensure there's a season at all. It is also widely used to top up the snow on runs that get icy during the season or lose snow cover due to wind or melt, or to build features like snow parks and jumps for competitions.

In other parts of the world, snowmaking has made skiing possible in places it previously didn't exist, like in Beijing, host of the 2022 Winter Olympics where 100 per cent of the snow was artificial, and Saudi Arabia, a country that just recently won the hosting rights to the 2029 Asian Winter Games (the "sustainability" arguments around this event are weak at best). In Dubai, Manchester, and New Jersey, there are indoor ski resorts popping up, made possible by snowmaking technology. I'll spare you the lecture on how terrible these indoor facilities are from an environmental standpoint, but suffice to say, they're bad – the energy use is through the roof.

For all the business stability snowmaking has provided, it isn't a simple technology to pull together, at least not at scale. Several features must come together for snowmaking to work: access to water, pumping capacity (a way of getting the water up the mountain), sustained cold temperatures below -2 degrees Celsius (28 Fahrenheit), automation systems that read weather data and allow for snowmaking to start and stop automatically at the most opportune moments, and a lot of cash. The biggest costs, once the initial infrastructure is in place, are labor and electricity. Only specialists can operate the snowmaking systems, and they're expensive to hire. And pumping water uphill and into snow guns is energy-intensive. The cost of producing snow for one hectare (10,000 square meters) with 6 inches (150mm) of coverage can vary between $1,000 and $2,000 (£800–£1,600) and it takes between one and six hours to cover the area.

The tricky part of snowmaking, and the part hydrologist Professor Carmen de Jong at Université de Strasbourg in France can't look away from, is the water costs. Her research demonstrates that it takes 3,000–6,000 cubic meters of water (or 660,000 to over a million gallons) to produce a hectare of snow, but that's assuming all the conditions are perfect. In some cases, where it's particularly dry, or warm, or windy, that number can swell to 12,000 cubic meters of water.

On a call in February 2023, de Jong explained her concerns to me this way:

> The industry experts will always say their snow guns are getting better and using less energy and less water. And yeah, you can use less energy, but you can't use less water because it's a physical law. You need a certain amount of water to produce a certain amount of snow, you can't change those ratios.

Industry-wide, water is the main ecological problem. The industry leaders will typically say there's enough water available locally from rainfall, snowmelt, and rivers to fill their reservoirs for snow-making. But according to de Jong, that's never the case. From a hydrological point of view, these catchments are tiny – we're talking about 120 square kilometers (46 square miles), and you cannot fill a reservoir of 400,000 cubic meters or even 100,000 cubic meters with such a small catchment area. De Jong tells me, "That's physically not possible, unless there is a huge river flowing through it. But most of the Alps are in dry regions, same with the Rockies. And yes, there is rainfall, but that will only give you maybe 5 per cent of the water needed. So they always have to get it from somewhere else."

The way ski resorts draw water for snowmaking is complicated; they source it from all over. Some resorts get water from the drinking-water network, or from other catchments, or from downslope sources. And if there is enough water available locally, often the streams are too small, and that water is being diverted from other needs downstream, such as farming or drinking water.

Over time, the use of reservoirs has become viewed as an ideal solution to the lack of water because it creates a pool from which the resort can draw a lot of water quickly to produce all the snow in one go at the beginning of the season. "But that's not natural," says de Jong, "that doesn't go with the seasons or with nature or with climate." Since expectations of good early-season conditions and December opening days have grown, the resorts need tremendous amounts of water all in one go. "What I am very concerned about is that increasingly, the local streams are not sufficient anymore, and they've tapped all the other sources. Recently, some resorts have started pumping groundwater. And I think that will be the end of the story."

Pumping groundwater is dangerous, expensive, and unsustainable. Dangerous, since it can cause conflicts with drinking-water supply. For example, in Bavaria, water is pumped from the valley bottom from a depth of 50 meters (165ft) below the surface. At the same time, local communes pump groundwater for drinking water. During droughts, conflicts with drinking water are likely to occur, and have already occurred during normal winter conditions. Using groundwater is expensive since it first has to be pumped from great depths to the surface, then transported to the bottom of the ski resorts, then pumped all the way from the valley bottom to the top of the mountain, often 600 to 1,000 meters up (2,000–3,000ft) or even higher. In smaller ski resorts such as Brauneck, Bavaria, the costs for this already reach €1 million ($1.1 million, or £860,000) per year. Using groundwater is also unsustainable because the water is transported uphill, causing high CO_2 emissions. What's more, it has nothing to do with natural snow or local water resources.

Atle Skårdal was a renowned World Cup skier in the downhill and super G series in his day. He competed in the Calgary and Lillehammer Olympic Games in 1988 and 1994. In his racing career, he won several big races including the Kitzbühel Stein and Val Gardena in the 1990s, taking home the World Cup Series championship in 1996 in Super G. He now serves as the Alpine Technical Director at the International Ski Federation (FIS). Over a call in the summer of 2021, he told me:

> Snow conditions are the most important topic for skiing races. If you have good snow conditions, you have a better race in terms of racing and safety. In the last 30 years, snow quality and snow preparation has improved tremendously. But it's harder now to maintain the conditions because the skiers are much better now, turning faster, moving more snow as they ski, and so it's hard to maintain. If you look at the videos from old ski competitions you can see conditions that were not at all acceptable by today's standards.

His job is also, unsurprisingly, made harder by climate change.

> It is a big concern for the mountains at lower altitude being hit harder now by lower temperatures, and even extreme weather

change in the short term. It can go from below freezing to warmer temperatures in just a few hours. Everybody in the industry is looking for the best possible opportunities to produce snow. The companies interested in snowmaking are coming up with good solutions. They can now make snow in plus temperatures conditions, which is energy intensive, but it's possible.

Possible, yes. Sustainable? Less so.

Other recent developments in snow management are snow farming and snow transport. Snow farming, which has become a trend over the last ten years, is an attractive option financially because it involves producing a lot of snow during dry cold weather in the winter, and storing that snow under tarps throughout the summer for use in the fall, instead of producing very expensive snow in marginal conditions and with marginal outcomes in October and November. Snow farming is what has allowed for fall training to continue in many cases, which is important for the athletes. Snow transport, on the other hand, is less ideal from an environmental standpoint and essentially involves grabbing snow from one place and transporting it in a truck or by helicopter to somewhere else. This was done to "save" skiing at the Vancouver 2010 Winter Games, and it was not a good look.

In 2023, Gstaad in Switzerland ran nine helicopter roundtrips from the mountain tops to parts of the resort towns at lower altitude in an attempt to dump enough snow to link two resorts. The snow transport caught the attention of global media and caused uproar among local environmentalists. And it didn't even work. In a statement from the city's tourism office for the *Daily Telegraph*, officials stated:

In this exceptional weather situation, snow was transported on this one occasion by helicopter from a nearby snow depot to guarantee the slope connection between Zweisimmen and the Saanenland.

This emergency measure was ecologically insensitive. It also proved to be unsuitable from a technical point of view. The transports by helicopter were therefore immediately stopped. No further flights are planned.

Searching for an answer

I can hear you asking: so what exactly is the solution? How do we run ski resorts better? And the answer isn't good. There is no real sustainable option; some resorts will just have to close. Those with sufficient natural snow can continue, and we can expect to see more snow farming in future, to preserve whatever snow falls for as long as possible.

Developers know all this: the high costs, the low water availability, the shifting winter season. Still, they see potential for snow sports and the tourism industry that surrounds it to continue making money, if only they can find more snow. So they scour the Rockies and the Alps for new powder. In the 1990s, one Japanese company hired Oberto Oberti, an Italian-born Vancouver-based resort architect, to find a location for a high-altitude resort.

Oberti found promise in British Columbia's Purcell Mountains – a range of pristine wilderness in a pocket of British Columbia between Calgary and Vancouver, at the headwaters of the Columbia River. Excited, the company drew up plans for a massive resort that would offer 5,600 feet of natural snow-covered slopes serviced by more than 20 lifts, accessing four mega-glaciers. In the summer, the glaciers would provide 2,300 feet of piste on natural snow. Truly a remarkable proposition, as it would become the only ski resort in Canada to offer full-summer ski schools. They would call it Jumbo Glacier Resort, after the Jumbo glacier on the mountain, and it was to become the biggest ski resort in North America.

Almost immediately, the project was met with criticism and plagued with controversy. Kathryn Teneese is the Chair of the Ktunaxa Nation Council. Her people have lived in the Purcell Mountains for over 10,000 years, and consider the Jumbo sacred.

When I interviewed Chief Teneese for my podcast in 2020, she explained that her people weren't concerned with the skiing itself. If people could come for the day, ski on natural snow – with no artificial supplement, as promised in the development proposal – and go home, that might be fine. In fact, they were happy for people to come visit and explore the beautiful area.

The bigger concern, she explained, was the tourism impact: the hotels, restaurants, grocery stores, and mini-city that would be developed at the base of the resort. All this extra traffic and pollution would pose a threat to the grizzly bear population and other animals who are drawn to human food and thus more likely to be at risk in a high-traffic development.

But the Ktunaxa's legal claim went even deeper than that: Teneese explained that her main argument against the resort was the infringement on religious freedoms. These mountains were sacred to her people, and not to be trampled. Any development on the site would damage it so badly that it would be impossible to experience the spiritual connection in the same way.

The case went all the way to the Supreme Court of Canada, and the Ktunaxa lost, despite having support from local environmental groups, residents of the local Kootenay region, and the federal government. Other religious groups also supported the case, arguing that the Ktunaxa's religious rights should exist alongside all others, as enshrined in Canada's Charter of Rights and Freedoms.

Ultimately, the ski resort development project was shut down by Canada's goal to protect 1 per cent of lands through a new designation called Indigenous Protected and Conserved Areas. The Ktunaxa put together an application to be part of this initiative, and with help from a government staffer in the British Columbia government, they were successful. Finally, after 30 years of battling to protect their area, the land became protected under that designation and the project to build Jumbo Glacier Resort died. The result was welcomed by the Ktunaxa's neighbors from across the region.

A similar challenge is emerging in Arizona, where Snowbowl resort has proposed a massive expansion that is being challenged by local Indigenous nations. The nations fighting Snowbowl's existence since the 1930s include Pueblo of Acoma; Fort McDowell Yavapai; Havasupai; Hopi; Hualapai; Navajo; San Carlos Apache; San Juan Southern Paiute; Tonto Apache; White Mountain Apache; Yavapai Apache; Yavapai Prescott; and Pueblo of Zuni. Now the Indigenous nations are saying the resort's expansion plans will worsen the already disruptive effect of the resort on the natural environment and their spiritual connection to the mountain region.

One claim that is particularly jarring is that the resort's snowmaking activities cause a public nuisance. Specifically, the resort uses treated waste water to make snow. This is widely touted as a "sustainable solution", and the only available source of water given the arid conditions in the region. However, the practice of snowmaking with reclaimed water has poisoned the area's water table. Signs at the mountain warn skiers against drinking the snowmaking water, because of the bacteria it contains. But it's also not ideal for the animals that will drink that water as it melts, impacting local wildlife.

A Chinese ski development had a similar deleterious effect on nature. To prepare an alpine ski area for the 2022 Beijing Winter Olympics, the core area of Songshan National Nature Reserve was torn up. Originally founded in 1985, the reserve had dense forests and alpine meadows that housed several protected species including the golden eagle and rare Shanxi orchids. The location came as a surprise to local conservationists, considering Chinese law had previously prohibited entry to the reserve's core area except for scientific research with government approval. To accommodate the ski run, the boundaries of the reserve were redrawn. Nonetheless, the damage was done and biologists expressed concern to the media that the alpine meadows in particular would not be fully restored.

As it stands, we've run out of new frontier. And, it can be argued, snow sports have already surpassed the reasonable limits

to energy, water, and land use. This is where it stops. There are
no new natural spaces to explore for new snow, not without
compromising important wild lands and Indigenous territories.

We're now left to make a difficult decision: do we squeeze
these places for the last drops of tourism money and fun for
the few people who can afford it, at the expense of nature? Or
do we step back, stop further expansion, and pivot toward less
damaging activities more aligned with conservation principles
and Indigenous demands?

Auden Schendler, Vice President of Sustainability at Aspen
Snow Company, was quick to tell me that they're working with
the most sustainable technology available, but there's not much
more they can do operationally to get this right. So Aspen has
pivoted to a different approach: political advocacy.

> I studied climate in university, and so I got into this role and we
> did stuff. And we kept asking ourselves, is it enough? And the
> answer is, of course, no it's not enough. We're doing all kinds of
> work but ultimately it's changing the lightbulbs and making small
> edits here and there, and that amounted to about 10 per cent.
>
> We started rethinking our approach and the role of business,
> and departed from the traditional approach of just reducing the
> carbon footprint in our operations. We're looking at having a
> public impact through new and creative ways: writing op-eds,
> attending conferences, producing thought leadership. And is
> that working? We're obviously failing horrifically on climate,
> and the ski industry is doomed, so in that sense, no. Most
> businesses still follow a classic approach of reducing emissions
> but not fast enough, so no. But in another sense, it is working
> for us because we're doing things that are meaningful, and
> manifestly, they have done well.

Aspen has been successfully disrupting the way the ski industry
engages with sustainability. Auden told me the story of how a
few years ago, the resort was trying to cut their carbon footprint,

but it kept going up, even though they did everything right on paper: installed green boilers, adopted green energy, switched the lightbulbs, more energy-efficient infrastructure. But it turned out, the company they bought their energy from – their utility provider – had bought a coal plant. All of a sudden, a high percentage of the energy Aspen was consuming was coming from coal, instead of other slightly better sources. So Auden and his team started intervening in local elections and municipal boards to reduce that utility's power in the region. It was a slow roll, but over 15 years they were successful. By contrast, Vail Resorts declined to participate in that work because it was difficult and controversial, and they're privately held.

––––––––––

I'll admit, I'm part of this problem. As the esteemed American poet Taylor Swift would say, "It's me, hi, I'm the problem, it's me."

It's uncomfortable to grapple with that. It's complicated to grieve the sport I love and to wrestle with the idea of not skiing anymore, while holding space for those who are suffering way more severe losses due to climate change.

Fortunately, there are ways to ski without causing all the damage that comes with water pulling and energy use: backcountry skiing and ski mountaineering. Both have picked up significant participant bases in recent years, with ski mountaineering entering the Olympic program in Milano-Cortina 2026. It's not quite the same, and it takes far more effort – hiking up the mountain for several hours, only to ski down in 20 minutes, is not my idea of a good time, but I can understand why some love it. It's a way to keep the skis in play in areas that have natural snow, without relying on artificial snow and copious energy supplies.

We can also pressure our ski resorts to do better: use as little water as possible, even if it means slightly poorer conditions during snow droughts. Farming snow at the end of the season would protect as much of it as possible, so we aren't starting from scratch the next season. And we should join the advocacy groups that are working to get more attention from government officials to curb emissions fast, so we can preserve what's left of the sport.*

---

*A side-note on how hard it is to talk about this. The interview with Professor de Jong broke my snow-loving heart, and I told her as much. She explained that she also used to love skiing, that her family loves to ski, and that doing her research has cost her more than just the sport. A few years ago, local ski areas, tourism boards, and other municipal authorities put pressure on her university to shut down her research because it was bad for the image of the region. The university didn't support her, and tried to get rid of her. This led to a lawsuit that spanned several years of court dates at just about every level of the French legal system. Ultimately, she won, but the damages (stress, finances, job insecurity, etc.) were immense. All this to say: it's hard to tell the truth about the ski sector, because it's so politically popular in the regions where it exists. Those who try to speak up get blackballed by the industry and the powerful elites who benefit from the tourism dollars it brings in.

CHAPTER TEN

THIN ICE

It's said that we don't appreciate what we have until it's gone. Well, what about when it's on its way out?

Away from the slopes, in communities across Canada, the US, and northern Europe, ice is the preferred winter surface. In Canada, few places evade the months-long snowy winters. Pond hockey and skiing are deeply entrenched parts of "Canadiana" folklore, and parts we'll be devastated to lose.

I learned to skate a week after my second birthday. I don't remember it, obviously, but there are pictures of me as a toddler, pushing a red plastic sled around an outdoor rink in Toronto's downtown area in December of 1995. My dad taught me, with my baby sister – just days old – strapped to his body in a carrier. Skates are heavy on toddlers' feet, so you end up just gliding over the surface as a parent pushes you. Like I said, I don't remember this, but having watched younger family members go through this process in recent years, it seems like a ton of fun to glide around and discover the ice.

My dad was keen to see my sister and me learn to skate young. He had learned to skate before he could remember on an outdoor rink in Montreal, wearing hand-me-down gear from one of his many older brothers. My grandfather would also have learned to skate in early childhood at an outdoor rink near his home. My great-grandfather would have done the same.

Our story isn't unique. In many ways, learning to skate young
is a rite of passage in Canada. It's also a big part of our (superior)
hockey culture (sorry, Team USA and everyone else).

Colin Wilson played in the National Hockey League (NHL)
for 11 seasons with the Nashville Predators and the Colorado
Avalanche. A third-generation hockey player, he is the son of
Carey Wilson, who played in the NHL for 13 years, and grandson
of Jerry Wilson, who played for the Montreal Canadians for one
season in the 1950s. Colin was born in Connecticut while his
dad played for the New York Rangers, but he spent most of his
childhood in Winnipeg, shuttling between arenas or skating on
the pond behind his house. He credits access to outdoor skating
with Canada's success in the sport, which has (mostly) dominated
on the world stage for the last few decades.*

I think outdoor rinks are a big part of why Canadian players
are so good. Obviously, there's a culture around it in Canada,
but at the same time, it's just more economically viable to get
better at hockey if you have access to outdoor rinks. You know,
it costs a lot of money to go to indoor rinks for Americans
and Canadians, ice time is expensive. And you usually don't
get ice time to yourself. But with outdoor rinks, you just have
free access to it all the time, you can work hard whenever you
want and as much as you want. And that's what I did. I would
go out to my backyard, and practice. So I got it, I understood
pretty young that I could go out and play as much as I wanted
and get better. And I just took advantage of that. Like if you're
a basketball player, you can go and just shoot hoops if you want.
We can go and skate.

_____

*I should note that Canada has dominated the women's game for as long
as I've been alive.

Jonathan Toews, current captain of the Chicago Blackhawks, also grew up in Winnipeg with his own outdoor rink. Wilson could see it from the highway. And Ryan Garbutt, also a Winnipeg kid who grew up to play in the NHL, lived in a house that backed on to the same pond as Wilson. They all took advantage of the extra playing time and went far in the sport.

Since retiring from professional hockey, Wilson has turned his attention to climate solutions, noting an interest in climate tech and long-term energy storage. He's seeing the impacts of climate change on the news, but he also knows the impacts are hitting close to home in the hockey community. "Ya, I worry about it," he shared with me. "As a kid, it's all I wanted to do. I wanted to be on the ice. Now, when I go outside in New Jersey and it's dark and cold, it feels like home. But it's sad to not see the outdoor rinks. It just doesn't get cold enough here."

That wasn't always the case. Back in the early twentieth century, there were many outdoor rinks in New Jersey. Princeton, one of the founding members of the first intercollegiate hockey league in 1899, played outdoors on a brook near campus. In 1907, a better facility was constructed on Lake Carnegie – also outdoors – and the team went on to win their first championship that year. They won again in 1910, 1912 and 1914, spurring interest in the sport across the state. By the 1960s and 1970s, pond hockey tournaments hosted in New Jersey would attract dozens of teams. Those no longer exist.

Rink Watch, a citizen science initiative out of the University of Waterloo in Canada, has asked families with backyard rinks to send in regular reports of the ice and weather conditions through the winter season, for the past several years. Launched in 2013, they've collected a decade of data on more than 1,400 outdoor rinks. Their most recent report on the 2021/22 skating season saw the hardest conditions in the northeastern US, where the start date of the season was January 10 or later.

This isn't a future climate change problem, but an ongoing one. In addition to a later start to the season, the number of viable skating days on outdoor rinks across North America has shrunk by 20 to 45 days since the 1970s, depending on where you are. The trend is expected to continue into the mid-century and beyond. Considering the costs of maintaining an outdoor rink and the maintenance that goes into ensuring a smooth surface, many families are choosing to skip it.

In 2023, the Rideau Canal in Ottawa – which holds the Guinness World Record for the longest naturally frozen ice rink and welcomes over a million visitors each year – didn't open. For the canal to freeze, temperatures must stay in the range of -10 to -20 Celsius (14 to -4 Fahrenheit) for at least a week. But the weather just wouldn't cooperate. The city was forced to scale back outdoor activities for its Winterlude festival, which runs annually for two weeks in February, while hotel association president Steve Ball told Agence France Presse that "bookings are way down". In a cruel twist of fate, and a great demonstration of weather extremes typical of a climate-changed world, the first weekend of Winterlude saw temperatures too cold to hold any events, so they were canceled. Too warm for skating, then too cold for any outdoor play at all.

Pond hockey tournaments, a relic of Canadian and American folklore, are also on the brink. In February 2021, a tractor fell through the ice at Mirror Lake in Upstate New York as the driver was clearing the ice for the Can-Am Pond Hockey Tournament. In 2022, the annual New England Pond Hockey Tournament was canceled when warm temperatures, rain, sleet, and snow caused cracks and springs to open up, making the ice unplayable. A string of cancellations followed through 2022 and 2023, which saw dozens of pond hockey tournaments across eastern Canada and the northern US get canceled.

Some tournaments that managed to hang on had to be moved off natural bodies of water, onto land. For example, in Sudbury, Ontario, the annual Pond Hockey Festival on the Rock has permanently shifted from its original home of Ramsey Lake to the shores, where artificial outdoor rinks were set up. Some games were also moved indoors, to ensure there would be playing opportunities regardless of the weather.

In Finland, hockey lover Svante Suominen launched the non-profit Save Pond Hockey with Canadian Steve Baynes after the duo noticed their skating season was shifting due to climate change. Not wanting to sit on the sidelines, the pair set out to promote climate action through a series of pond hockey tournaments. They hosted the first tournament in February 2015 on artificial ice, because already the natural ice was not strong enough to support skaters. In 2016 they grew the event, and have since hosted tournaments in Finland, Denmark, and the US. Their tournaments have raised over €80,000 ($88,000 or £70,000) for renewable energy campaigns, marsh restoration efforts, and a range of local climate campaigns in the places where the events take place.

Elfstedentocht

In the Netherlands, speed skating is the winter sport of choice. Five million Dutch people – roughly one-third of the population – own a pair of ice skates, and a million people regularly participate. The country boasts 400 indoor rinks and the longest recorded history of skating. According to historian Marnix Koolhaas, skating became commonplace in Dutch culture in the Middle Ages due to an increase in wealth, a drop in metal prices, and a string of harsh winters. For those who didn't own horses, skating was the most efficient, fast, and egalitarian way to get around, and remained that way into the nineteenth century. This cultural focus has yielded incredible results on the world stage; the country has amassed the most successful Olympic record in speed skating: 121 medals, among them 42 gold.

At the center of the Dutch skating culture is the Elfstedentocht – a 200-kilometer (125-mile) race through 11 towns in Friesland that is scheduled to take place each winter, ice conditions permitting. The Elfstedentocht started informally in the mid-nineteenth century with people skating along the country's canals from town to town to visit family and friends. The race was codified and established as a competition by W.J.H. Mulier, a sports pioneer of sorts who also introduced soccer, hockey, and athletics to the Netherlands, and who had fallen in love with the event when he participated in the informal 1890 race. Together with the Frisian Ice Union, Mulier organized the first formal race in January 1909 with 23 participants. The next year, the Society of Frisian Eleven Cities formed to become the permanent organizing body of the event. Since 1909 the Elfstedentocht has been held only 15 times, growing from its initial 23 participants to 3,000 in 1940 and 16,000 in 1997 – the last time it was held. And yet it remains iconic in Dutch folklore.

There's no set schedule for the Elfstedentocht; it only happens when the ice is sufficiently thick, which organizers define as at least 15 centimeters (6 inches) in depth throughout. But the impromptu nature of the race is a big part of what makes it so exciting. Once the organizers decide that the conditions are right for an Elfstedentocht, plans are made, volunteers are called in from the race organization's 32,000-person membership database, and an announcement is made on television. The race is held within 48 hours of the official announcement, so skaters don't have much of a heads-up. The start time is 5:30 in the morning, and skaters must complete the 200km skate before midnight that same day.

Regardless of the fact that there has not been a race since 1997, it's been planned each year, just in case the weather cooperates. With just 48 hours to pull the event together, the plans have to be revised and updated, ready to deploy quickly. According to Wiebe Wieling, Chairman of the Elfstedentocht, he spends one day a week all year round on planning, except July and August. He leads a board of ten members who also support the planning, with support from 100 part-time volunteers. The plans are complete by December 1. But despite the longest dry spell in history (the next longest wait was 22 years between 1963 and 1985), organizers and skaters alike are reluctant to give up hope of a future event, fighting off climate concerns.

Every winter, Dutch skaters chase ice, watching the weather and sharpening their skates in anticipation of getting on the ice as quickly as they can once it's ready, to ensure as long a season as possible. The culture around skating is communal – skaters won't get on the ice until it's firm enough to hold all the people who might want to participate. In years with poor ice conditions, nobody will skate. It's viewed as a natural feature available to "all of us or none of us". To manage this, in each municipality, an official *ijsmeester* (ice master) routinely tests the ice and uses a flag system to indicate when it is safe for skaters. Of course, there are always a few exceptions who ignore these customs, but they are viewed as irresponsible and egotistical by their peers. When it looks like some skating might be on the horizon, the Dutch are ready.

Ecological nostalgia is defined by Dr. Olivia Angé and Dr. David Berliner as "longing for past forms of life in earthly environments" – an anthropological concept similar to solastalgia, which was introduced in the last chapter. Anne Veere Hoogbergen studied ecological nostalgia among Dutch skaters and found different responses among younger skaters who'd never participated in the Elfstedentocht, and those older skaters who remembered the event and had participated. She writes: "One of my most prevalent findings during my fieldwork is that although the older generation often worried about the ice-winters becoming scarcer, they remained optimistically hopeful about the Tour. They perceived the long wait for the Tour – now more than 24 years – as an inherent part of the Tour's uniqueness."

Veere Hoogbergen notes that these older skaters are ignoring the fact that during the 1970s, chemical discharge with cooling effects was dumped into the Frisian canals by power and industrial plants in the area. This process has now been stopped. In those years, without the chemicals, the canals would have needed temperatures at least 4 degrees colder than normal to freeze to the required 15cm (6in) thickness. So, she argues, the older generation may be holding on to "a distorted or an outdated idea of the Tour's rarity".

However, while older participants do not seem to be losing hope for another Elfstedentocht, they seem to be worried about – and nostalgic toward – ice-skating culture itself. The two concerns that were raised most frequently by this generation were a loss of knowledge and a decreasing interest in skating. The latter would have immediate and significant negative impacts on speed skating as the top winter sport.

Among the younger generation of Dutch skaters, their lack of experience with outdoor skating has led them not to be nostalgic for the loss of the activity, but anxious about losing a piece of national heritage in the Elfstedentocht. Veere Hoogbergen writes that "This manifests itself correspondingly in how people seek to keep their heritage alive. While the older generation is more focused on passing on this heritage by acquainting the youngsters with ice-skating by means of educational programmes, fieldtrips and storytelling, the younger generation directs its attention to gathering and collecting the stories in order to make the intangible tangible."

Sliding into sled-head

In some sports, concerns that poor ice conditions may lead to increased or more severe injuries are taking hold, though the science has yet to catch up with the lived experiences and testimonies of athletes. Bobsleigh, skeleton, and luge – collectively, the "sliding sports" – have a common challenge: sled-head, a cutesy name given to a very scary phenomenon. In sliding sports, athletes ride at extraordinary speeds down tracks of ice. The high speeds, vibrations along the track, and gravitational forces in the big turns are enough to turn any layman's stomach, but they also might be causing a set of symptoms not unlike those of a minor concussion: dizziness, fatigue, lack of focus, and brain fog.

Sledding is inherently brain-rattling. In even the smoothest conditions, there is considerable risk of suffering sled-head symptoms. Full-blown concussions are also a common occurrence, affecting 13 to 18 per cent of all elite sledding athletes. Most concussions are registered after a crash, as there's a clear indicator and visual signs that something went wrong. But with more subtle symptoms of sled-head, it can be tricky to tell what's worthy of concern and what's not.

An athlete in bobsleigh, skeleton, or luge will go down a track a few times in a practice, and sometimes practice twice a day. They wear helmets, and luge athletes also wear chin straps, to keep their heads safe as they travel down the track. But brains are soft, suspended inside the skull. Even with a helmet on, it's possible to be concussed, as the brain hits the edges of the skull on sharp turns or hard stops. Gravitational forces, or G-forces, are the measure of how much power gravity exerts on an object – or in this case, a body – and some research has been conducted to estimate how much G-force needs to be exerted on the human head to cause a concussion. While it can be hard to pinpoint the precise G-force that meets that threshold in different sports, one study found that most diagnosed concussions fall in the arena of 85–95G, with some registering after an impact as low as 60G.

For reference, walking down the street will be around 1G. A roller-coaster typically registers in the range of 4G to 10G. When Canadian physics professor and skeleton athlete Dr. Alexis Morris strapped an accelerometer to his helmet on a run of the Whistler Olympic track, he recorded speeds of 70 to 80 miles per hour, which is pretty standard on that track, and a high of 84.5G in one particularly sharp corner. That spike only happened for a fraction of a second, but that can be enough to cause damage, and several smaller spikes in the 40–60G range happened along the course. What's more, skeleton athletes don't go as fast as lugers or bobsledders.

There's more than speed and G-forces contributing to sled-head. There's also the vibrations. Long-term exposure to vibrations can be bad for the body, shaking up organs and impacting the body's capacity to perform basic functions. The International Standards Organization (ISO) has developed guidelines for how much vibration a person should encounter in a year. In one study, Dr. Peter McCarthy at the University of South Wales attached vibration-detection devices to the helmets, sleds, and backs of skeleton athletes to read how much vibration they are exposed to on a run. Preliminary results showed that doing just one run on a skeleton sled could put you over the ISO's annual limit. The consequences of this are hard to determine and require far more research, but the baseline finding is bad: sliding athletes experience a lot of vibration, which may increase their risk of internal injury or micro-concussions.

Normalizing concussive-like symptoms with a cheeky term like sled-head is dangerous, according to brain experts, because it can lead athletes to believe it's not a big deal. McCarthy told me his biggest concern is that if head injuries aren't properly identified and treated, it could lead to long-term damage. And if the symptoms are chalked up to "normal" side-effects of the sport, that makes it easy to look away and not take care of the athlete.

"The big issue in sport injuries is always the same one, protecting the athletes from themselves. They just want to play, they're competitive, and it's fun. There's nobody there to protect the person. If you've got a broken leg, then yeah, they're all out there protecting you. But if you show signs of concussion or dizziness, it's just not viewed the same way. And that could be costing lives." He went on to explain that he has presented his research findings to the board at the International Bobsleigh and Skeleton Federation, but they fell on deaf ears. "Sliding sports could be killing people slowly, but the federations don't want to support the research that would help us empirically find out, because it could be bad for business."

Sliding sports have seen high rates of concussions and post-concussive symptoms like depression, anxiety, and in some cases, suicide. In 2014, Travis Bell, who competed for the

United States in the 1990s, took his life at age 42. An Olympic medalist, Bill Schuffenhauer, attempted suicide in 2016 but was saved by his girlfriend. The following year, Steven Holcomb, who piloted the sled known as the Night Train at the Vancouver 2010 Olympics for the US, winning the country's first bobsled medal in 62 years, died alone of an overdose. This is a sport with fewer than 500 athletes competing internationally, making these rising rates of suicide alarming.

While it's hard to pin down the cause of the suicides, each of these athletes had experienced head injuries while competing in their sport. One more recent case adds to the body of evidence that concussions in sport might be part of the problem: Pavle Jovanovic, a former bobsledder for Team USA, was already experiencing the shakes and tremors often associated with Parkinson's disease in his early 40s. He died by suicide in 2020. A year later, Dr. Ann McKee, a neuropathologist and the director of Boston University's C.T.E. Center, who has discovered evidence of chronic traumatic encephalopathy (CTE) in the donated brains of dozens of football players, found evidence of the disease in Jovanovic's brain.

A meta-analysis of studies on sled-head found only a few empirical studies on the phenomenon, but based on available evidence, it appears that ice quality is a factor. The researchers were able to identify this based on studies that examined sled-head incidences at several different tracks that had different ice smoothness and surface quality. An uneven ice surface leads to a bumpier ride, which can lead to more concussions.

Hannah Campbell-Pegg, two-time Olympian in luge from Australia and current head of the National Luge Federation, nodded when I shared that finding. "That sounds right. If the track is in bad condition, at the end of the day when everybody has been down it, or just on some tracks that are harder to maintain, you feel it. The sled-head is worse at the end of the day." She's currently pursuing a PhD at the University of Canberra, researching how sliding sports are being impacted by climate change, but she hypothesizes there will be fewer functioning tracks in a warmer future, and those that stay open might become harder and more expensive to maintain.

Four-time Olympian in bobsleigh Chris Spring started his career with the Australian team before joining the Canadians after his first Olympics. He confessed that it's hard to talk about head injuries and mental health, and most athletes in the sport don't want to, because it's uncomfortable. He's known a few teammates and competitors over the years who've suffered with depressive symptoms, and some have died by suicide. While it's impossible to assert causation between the sport, sled-head, and these tragic outcomes without an autopsy report – which are often denied by families of the deceased – it's enough to make Spring think hard about his future in the sport. "It's something that I definitely think about, and moving forward, it may be one of the reasons why I'll choose to retire. I can't keep doing this forever. And maybe I don't want to do this forever because of those reasons, because of those risks."

An objective assessment of head injuries in the sliding sports has not yet been done. It may also be necessary to review incident reports at sliding tracks and on sliding teams, and include details on the track conditions in the incident reporting so sled-head, and the conditions on the track, can be monitored over time. I contacted the International Bobsleigh and Skeleton Federation for comment on sled-head issues, concussion protocols, and track conditions in a warming world, but they didn't respond.

———

Deteriorating ice conditions are threatening the cultures, economies, sport prowess, and health of athletes across the skating and sliding sports. With participation numbers already in decline over the past decade, it's hard to see how these sports will carve out a future in a warming world, aside from moving indoors completely.

The winter sports industry has been dubbed a canary in the coalmine by climate researchers. Each iteration of the IPCC's global climate reports has explicitly mentioned the winter sport tourism sector as being at risk. But here again, the worst outcomes can be curbed, and in the process, jobs can be created, injuries avoided, and tourism economies boosted. It's not just about sports, it's about the winter season and winter culture as a whole.

CHAPTER ELEVEN

AT A DISADVANTAGE

Like most social problems, climate change does not impact all communities and nations equally.

Globally, the impacts of climate change are hurting historically exploited nations first and worst, both because their geographies leave them vulnerable to extreme weather and because their capacity to respond and adapt to climate change is limited. Most of this book so far has been about big sports competitions and impacts in Europe, North America, and Australia, because that's where I've done most of my research. These regions are also home to some of the richest sports leagues and most of the international sports federations, so there's a concentration of power in these richer nations. This is not only true in sports. The so-called "Global North" holds the balance of power in just about *every* aspect of the global capitalist system. This is not a design flaw but a feature of our current system.

Let me break that down a bit further. In the nineteenth century, as the Industrial Revolution saw the beginning of fossil fuel extraction for energy production, transport fuel, and manufacturing, many of the countries with the natural resources (minerals, metals, coal, fuel, agriculture) and labor required to make the Revolution possible were located in Africa, Central and South America, and Asia. Many of these countries were not independent at the time – they were colonies of the much (financially) richer and more powerful empires in the Europe. Across the colonies, lands were ravaged by colonizers and private capitalists based in the empires of Europe for natural resources, while the people in the colonies were put to work in manufacturing, mining, and agriculture, producing the resources that would be sold back to the empires.*

None of what I'm saying is new. It's all in the historical record, but it's important to tie these things together. The fossil fuel industry, which is responsible for more than two-thirds of global warming according to the Carbon Majors Report, was born out of this colonial project, which made it possible – acceptable, even – to plunder colonies for their resources.

As historically exploited nations slowly gained their independence politically through the twentieth century, their economies remained tied to their former colonizers as many extractive businesses were privatized by (mostly) white owners in the North. So resources and money continued to flow North, leaving newly independent countries to grapple with a limited capacity to grow their economies, and in exploitative economic relationships with their former colonizers. Countries that have experienced colonization were the source of many riches, without access to the rewards.

According to Carbon Brief, the United States has released more than 509 gigatonnes of carbon dioxide since 1850 and is responsible for the largest share of historical emissions – roughly 20 per cent. China is a relatively distant second, with 11 per cent, followed by Russia at 7 per cent. But these numbers can be deceiving. Two large post-colonial European nations, Germany and the UK, account for 4 per cent and 3 per cent of the global total respectively, but that doesn't include overseas emissions from their colonies. This last part is important. The way the historical totals are calculated, former colonizers are not held accountable for the emissions produced in their colonies. Meaning they

---

*Some of these products get sold back to the South as finished products at exorbitant prices, but that's another story.

get to benefit from all the riches produced there but shrug off responsibility for the emissions. That's some creative accounting.

All this is to say: some countries are more responsible than others for creating this global mess. And ironically, those countries happen to be some of the least impacted. As writer Mary Annaïse Heglar has repeatedly pointed out in her essays, the places with the most political power – the Washingtons, New Yorks, Londons, and Genevas of the world – are far removed from the frontlines of climate change. These cities are also home to the global media complex, meaning impacts in these countries receive disproportionately more coverage, while the challenges in the historically exploited nations (that are not responsible for climate change) go largely ignored. So, in this chapter, I want to shed light on how climate change impacts are coalescing and compounding in places like Kenya, Pakistan, and Fiji, reshaping the way sport is played in some parts of the world.

Drought and heat in Kenya

The drive from the tiny Eldoret airport to the town of Iten in the south-west corner of Kenya takes about an hour. It's a winding unlit road with few road signs: you need to know where you're going to get there. The town's population isn't known – there hasn't been a census in over a decade – but the local municipal authority estimates it around 56,000, up from 40,024 in 2009. Roughly 35 per cent live below the poverty line. And yet a sign on the only paved road into town calls this the Home of Champions, owing to its phenomenal athletic success.

This corner of Kenya has produced 14 men's and 9 women's Boston Marathon winners since 1991, who've brought home 22 and 14 wins, respectively. They've also won 13 of 18 gold medals in the 3000m steeplechase at the World Athletics Championships since the event was introduced in 1983. One local high school, St Patrick's, a squat building in the middle of Iten, has produced several world champions in track events. In

2019, World Athletics named the town a World Heritage Site for running, and it's easy to see why. The rich culture of running is everywhere in this place.

I rolled over groggily at 6am on a Saturday morning in our tiny guesthouse room. It was still dark out, but you could hear the town stirring awake. Humans, roosters, dogs, the occasional cow. Along the main road, the first few runners could be seen logging miles along red clay roads at an impressively fast pace. By 6:45am, a runner or pack of runners was passing us every 30 seconds or so, gathering to stretch and chat with their coach at the end of a long 30km (18.5-mile) circuit.

I counted more than a dozen shops and small businesses with "champion" or "Olympic" in the title in the 1km (1,100yd) stretch from the town gate to the main commercial area, selling everything from shoes to hats to training plans. There are also signs advertising training centers: Olympic Champion Lornah Kiplagat's High Altitude Training Center was the original, but since its opening in 1999, a number of other private clubs have opened in town. These have become tourist attractions for the world's growing running community, drawing eager athletes from Europe and the Americas to test their mettle against the locals.

Nearby, in Kaptagat, a rural town less than one hour's drive from Iten, some of the fastest distance runners in the world live and train. These rural towns are similar. Life here is simple; there are few distractions, and at 2,400 meters (almost 8,000ft) above sea level, they offer near-perfect training conditions.

Many have speculated whether it's nature or nurture that accounts for the success of Kenyan runners. According to Professor Vincent Onywera, the deputy Vice-Chancellor for Research, Innovation and Outreach at KCA University, who has studied Kenyan runners most of his professional career, it's both.

I met with the professor in his Nairobi office on a sunny day in May. He explained to me how he has spent significant periods of time in Iten, studying the runners. His whole doctoral dissertation tackled the question of nature or nurture, and his subsequent 30-year career in pediatric physical activity research has continued the work, investigating young elite runners in the region. He calls the Kenyan runners the beneficiaries of a "phenomenon effect", explaining that champions are not just born with these gifts, they're tactfully trained to excel in the sport. It's a potent combination of good genes, a balanced and hearty diet, a supportive social environment, a psychology of winning, and an ideal high-altitude training ground.

Southwestern Kenya, where Iten and Kaptagat are located, is an agricultural corridor that produces crops of maize, beans, cassava, sorghum, peas, a variety of fruit, and livestock. The cost of foodstuff in the region is cheap, so it's possible to get enough food, and enough nutrients, to sustain a strong runner's diet on a low income. This is important as many of the country's most impressive runners come from modest backgrounds. Eliud Kipchoge, former world record-holder in marathon, grew up in a single-parent household. His mother was a teacher. In his teenage years, he made money delivering milk on his bike. Despite modest incomes, the availability of food was never an issue, even for a budding athlete who needed robust and nutritious meals, because local farms supplied food for the region at affordable prices. It was simple but nutritious.

Culturally and economically, the residents of small running towns recognize the benefits of the running culture and the tourism it attracts. In Iten, the municipality has drafted a Tourism Action Plan to "leverage upon its nationally and internationally recognized position as a world leader, as recently acknowledged through the IAAF (World Athletics) award". There's also, according to Professor Onywera, a growing awareness of the benefits of physical activity, and a growing number of parents who wish for their children to run. Joan Chelimo, a professional runner and multi-award-winning athlete, told me over a cup of tea that "families have begun to realize the potential for runners to make money on their winnings and through sponsorship. With this, parents have become more supportive of their children running."

By extension of the supportive social environment and the track record of success in the sport, athletes from Kenya have developed a psychology of winning. Professor Onywera explained it this way:

> When a Kenyan runner lines up at the start next to a European runner, the European runner will typically get worried because they'll be aware of the Kenyan success in the sport. The Kenyan runner, for their part, will expect to beat the European runner because they've been taught their whole lives that Kenyans are the superior runners. Success in running is viewed as a rite, anything less can be devastating.

This psychological starting point leads athletes to compete hard, sometimes through injury and adverse conditions, to chase that win. It makes for a strong foundation, albeit occasionally a problematic one.

Perhaps the most striking factor contributing to Kenyan runners' success, and the factor that foreign runners visit the country to experience, is the near-perfect training conditions. At 2,400m (almost 8,000ft) altitude, Iten and Kaptagat are ideally located for high-altitude training, which has been proven to increase the lung capacity of athletes. It also benefits from long, unpaved clay roads, a softer surface to run on compared to pavement, which allows the locals to log more miles without risking the lower extremity stress injuries so common elsewhere.

Slowly, though, the environment is changing in this powerhouse running country. Kenya sits along the equator, the region hardest hit by climate change. Throughout East Africa, the most obvious climate hazards so far are drought, extreme heat, floods in some parts, and associated biodiversity loss. Professor Onywera spelled out what those climate change impacts could mean for Kenyan runners: "I expect injuries to go up as droughts make clay roads harder than pavement and the impact on knees and hips takes its toll. It won't be something we see right away, it'll be something we see over time."

Local doctors are also aware of this risk and have started running observational studies to set a baseline on the injury profiles of runners in south-west Kenya. Dr. Mbarak Abeid, an orthopedic surgeon based in Eldoret – the nearest major hospital to Iten and Kaptagat – confirmed that so far, injury rates to lower extremities are not as high in this population as we might expect, owing to the softer surface of the clay, but that we might see this change over time as drought dries it out.

Drought is also impacting food crops. Kenya is in the middle of its worst drought in 40 years. In the parched north of the country, rivers are running dry and millions of livestock have perished due to lack of food. With much less irrigation, machinery, and fertilizer on hand to cushion the blow, large swaths of the country are experiencing a dwindling food supply. In 2022, the Kenyan government made the controversial decision to reverse the ban on genetically modified foods to alleviate food stress. And while the droughts aren't yet impacting the south-west region, where a majority of the elite athletes live and train, the ongoing crisis is certainly impacting the youth levels of the sport.

"The usual rains that used to come to maintain the environment are no longer coming. And so dry lands, droughts, food shortages, and sometimes fires that produce smoke, are impacting those areas, and impacting runners," said the President of Athletics Kenya, Jackson Tuwei. "If young athletes are hungry, they won't run. If the air quality is bad because it is dusty and smoky, they won't run. If there is no shade because trees have been cut down, and there's no water to drink or shower with, they won't run." Drought will cut off the pipeline of athletes in Kenya unless solutions are delivered soon.

Heat is only making things worse. Tuwei named this as his first concern for Kenyan athletes. "In Kenya, we don't have seasonal changes, so we have 12 hours of daylight and 12 hours of nighttime, basically all year. When we are running in the daytime it can be extremely hot, and so sometimes we have to change our programs to run in the evenings or very early in the morning. We have witnessed high temperatures here in recent years, forcing us to change our schedules for events." Training schedules are changing, too. The 4pm afternoon run, a staple of the two-a-day training regimen popular among the most decorated runners, is happening later, at 5 or 6pm.

Sleep is another casualty of heat. Onywera stressed that a healthy sleeping schedule is critical to runners' success, but that warmer nights across the country are shifting performance outcomes for some. "You can see the impacts of heat at nighttime. Concentration dips, motivation is low, energy levels are low. In the long term, if this keeps up, it could throw an athlete off their training regimen." The look on Onywera's face changed as he stopped to consider this further. He went quiet. "Yes," he continued, "sleep could be the most important loss."

Athletics Kenya is worried about how climate change might shape the future of its country, let alone its sport. And yet there is little a country like Kenya can do to improve its climate outcomes, as it contributes only 0.05 per cent of global emissions. Even if significant reductions in emissions are achieved, it will barely move the needle. Nevertheless, Athletics Kenya is committed to measuring and reporting on its carbon footprint, planting trees in rural areas to improve air quality and increase shade, and educating its top athletes to be spokespeople for climate change on the world stage.

"When our runners go out and do well," Tuwei said, "we want them to also talk about the environment because it is affecting us here, it is affecting them, and maybe people will listen."

Floods in Pakistan

Pakistan, another former British colony, is facing a different set of challenges.

In Pakistan, the summer of 2022 brought devastating floods that killed nearly 2,000 people and caused trillions of rupees in damage and loss. The floods were attributed to a toxic cocktail of glacier melt – Pakistan has more glaciers than any other place on Earth outside the polar regions – and heavy monsoons, both linked to climate change.

Ali Athar Nakash Brohi is the President of Mehran Football Club in Thatta, Pakistan, in the heavily impacted region of Sindh. His club was spared from the worst of the floods, which is why he was still online and able to talk to me. Nearby competitor clubs were less lucky. "Sindh was impacted more than other provinces, but Thatta got lucky. So we're giving our homes and shelters to people fleeing from the floods."

From the start of the floods, sports shut down in the region. The Mehran Football Club had dozens of families living on their grounds in makeshift shelters. "Some of our members who live in the villages outside Thatta lost everything. They're the ones that came here to stay on the grounds."

The flood took a significant toll on mental and physical health for those impacted and all the club members who rallied to support them.

> Our activities aren't going on and children are facing a situation where school is stopped, sports are stopped, and the mental health toll is a lot. Three months after the floods, the kids were really scared of floods. It was top of mind, so we try to distract them with small games. We believe play is a major part of child development and can be a coping mechanism to help them restore mental and physical health.

It took months for the displaced people to find permanent housing arrangements and for sports programming to resume. Even when it did, it was a slog. "It took many months to get our fields back in order and to start sports again. It was really challenging for us because everyone is completely disturbed, and everything we do is run by volunteers. Nobody has mental energy after floods."

In the weeks following the disaster, aid money poured in from foreign governments and charities. An estimated $9 billion (£7 billion) in relief funds reached the country. But with more than 30 million people displaced, that's only $300 (£240) per person – hardly enough to provide healthy food, clean water, housing, healthcare to cholera patients, and comfort to the victims, even if every dollar was used effectively. Yet, after the first two months, aid fatigue settled in and the initial outpouring of money ebbed.

The best the country could do was to have its most visible figures, cricket players, keep talking about the floods. Armbands were worn by professional athletes playing abroad, then special kits were created for the national cricket team. In the weeks after the event, the country's top athletes began donating their earnings to the flood victims in an effort to help, and the National Stadium allowed impacted locals to attend the England versus Pakistan cricket match for free in an effort to boost spirits.

It still wasn't enough. Six months after the floods, more than 10 million Pakistanis were still without clean water, and tens of thousands remain displaced, their homes having been completely destroyed in the floods.

Dr. Friederike Otto at Imperial College London calculated the likelihood that these floods were attributed to climate change. The short answer is that the flooding catastrophe had climate change written all over it. In a statement, Otto said,

> What we saw in Pakistan is exactly what climate projections have been predicting for years. It's also in line with historical records showing that heavy rainfall has dramatically increased in the region since humans started emitting large amounts of greenhouse gases into the atmosphere. And our own analysis also shows clearly that further warming will make these heavy rainfall episodes even more intense.

None of this is Pakistan's fault. The country barely emits 1 per cent of global emissions. Still, they're bearing the brunt of climate damage.

Losing homes, losing sports

In some places, sports are disappearing entirely due to rising tides. Low-lying Pacific islands are particularly at risk. Earlier in the book, I wrote about Tuvalu's beach volleyball team who have lost their beaches. In Fiji, the white sand in Namatakula, a small town of 600 residents on the Coral Coast of Viti Levu, is being swallowed by the ocean.

For reasons unknown, this small beach community has produced some of the best rugby players in Fiji's illustrious history in the sport. Lote Tuqiri, who represented Australia in both rugby league and rugby union, and Fiji in rugby league, and Tevita Kuridrani, who has 60 caps for Australia in international rugby, both took their first tackles here. Kuridrani told me,

> Rugby is a very important sport to our village community, and the beaches were the places where we used to play rugby as young kids and where the men would do their rugby training. It was a very special place and very important to us. With the beaches quickly eroding, it is impacting the community, especially with the rugby training program. Kids are growing up knowing they won't have the same experience and luxury of playing on the beaches as we used to do growing up. And that could impact a young kid's dream of being a rugby star.

Like runners in Kenya, rugby players in Fiji enjoy special status in the country's social fabric. Fiji is a rugby nation, with 80,000 registered players – more than any other country per capita – and the only two Olympic gold medals in the men's sevens game. Theirs is a quick and powerful style of play, which the national teams have repeatedly attributed to the logic of *vaku vanua*, "the way of the land".

Managed retreat is now being proposed to Namatakula residents as the primary adaptation option. The plan is to move away from the rising sea level and rebuild the town further inland to preserve the culture, local economy, and way of life. On the opposite side of Viti Levu, two towns have already pulled away from the coastline: Vunidogoloa and Narikoso. It's a complicated task and not all town residents are keen, but it may be the only option. Tropical cyclones Yasa and Ana hit the nation in quick succession at the beginning of 2021, laying waste to several towns. They barely had time to rebuild before Cyclone Cody came in hot just one year later.

In 2019, the Fijian government launched the world's first relocation fund for people displaced by climate. In 2020, New Zealand become the first international party to donate to the fund ($1.2 million, £900,000). This is part of a broader movement to calculate the toll that anthropogenic climate change has taken on developing nations, and to have fossil fuel-emitting rich nations compensate those most heavily impacted by the crisis and participate in funding recovery efforts. While not yet mainstream, loss and damage funding, initially proposed in the COP19 in Warsaw in 2013, has been at the top of small island nation states' negotiation agendas in UN meetings for the last decade. At COP27 in Egypt in 2022, a breakthrough agreement to provide loss and damage funding was agreed by the 190 countries in attendance. As I write this in 2023, however, no clear commitments have been made by rich countries regarding their contributions to the fund, or at least none that approach the scale of the true costs.

In 2022, the Fijian government formed a special taskforce to study a range of options for climate adaptation, and the resulting document – ominously titled Standard Operating Procedures for Planned Relocation – is perhaps the most detailed plan for managed retreat in the world. The taskforce identified 800 communities along shorelines that will need to be relocated, 45 of which are at high risk and will have to be moved in the next five to ten years. For the rugby players who train on Fiji's beaches, the loss will be deeply felt.

Loss and damage funding will be critical to survival in some places. Many developing countries on the frontlines of climate change are facing recurring disasters that they can't afford to bounce back from, let alone think of building up their economy to move forward. They're stuck in limbo. Even in a scenario where the worst climate impacts are avoided, public funding for other things, like higher education, sports, and the arts, will wane in the face of climbing debts.

At the Sport Positive Summit in October 2022, Julie Duffus, Senior Manager of Sustainability at the International Olympic Committee, told the audience that "I work with 206 countries around the world. And so many of them now are witnessing these impacts … by the time we get to 2030, we're looking at losing about 20 per cent of our Olympic nations. Literally, gone." I knew it was bad, but I didn't realize it was that bad. But of course, when you consider funding for sports as part of the financial picture for heavily-impacted nations, it's just not going to be there. Not when there are life-and-death crises happening on a regular basis.

CHAPTER TWELVE

PLAYING CATCH-UP

No sport will be immune from climate change. The hazards may not look or feel the same and the consequences will vary, but at some point, every sport – and everybody – will feel its impacts.

Acknowledging the problem is the first step. Most of the people I've interviewed about climate change in sports over the years didn't do much to prepare for the hazards they faced. It's not that they hadn't considered the possibility of a storm or a heatwave, they just assumed it would never actually happen to them, or thought they had more time before it did.

As the saying goes, the best offense is a good defense. We have to protect what we have, prepare for worst-case scenarios and insulate the most vulnerable populations from harsh impacts. Before we can move forward with curbing emissions in sports, we've got to play catch-up on adaptation.

When I bring this up to sports executives, the first thing they say is "okay, but that's expensive". They're right, of course. New equipment is expensive, so are new technologies and insurance plans, but failure to adapt could put sport out of business. The choice isn't to adapt or to carry on with business as usual. The choice is to adapt or to lose sport as we know it.

The good news is that not all adaptation strategies come with a hefty price tag. In addition to the obvious ones, such as new buildings, roofs on outdoor facilities, better land management technologies, weather radars, insurance plans, contingency funds, and so on, here are a few adaptation solutions that can preserve the way sports are played, and the venues we love, with a little creativity and flexibility.

Change the calendar

Sporting calendars are traditionally static. In the US, the American football schedule dictates the season begins in late August or early September, with Friday nights reserved for high school, Saturday for college, and Sunday and Monday for professional games. For association football, the season starts in August and runs through May with your favorite team playing a game every weekend. Marathon runners can expect races to be scheduled in the spring or fall, in the early morning, to avoid workday traffic and finish before the midday sun. The Summer Olympics are held in July and August, and the Winter Olympics are in February, with the Paralympics following two weeks later. Across the professional sports world, broadcast schedules dictate the timing of the most important and high-profile games. After all, broadcast is one of the biggest revenue drivers in the industry.

None of these traditional reasons for scheduling sports holds much water anymore. On-demand sports viewing has skyrocketed in recent years, especially with DAZN, Apple TV, Amazon Prime, and other streaming platforms entering the game. What's more, fans aren't working nine-to-five jobs like they used to. Due to COVID, work-from-home options and dynamic scheduling have changed the timing considerations for a significant swath of the working population in places like North America and Europe.

My colleague Dr. Jessica Murfree at the University of Cincinnati and I have spent the last year studying the precedent that's been set by COVID rescheduling efforts. Long story short: they were impressive. Sports went offline at the beginning of the pandemic, for obvious reasons, then came back on odd schedules to consistently strong viewer numbers. Sure, people had less going on during the pandemic and could watch more sports, but viewer numbers have climbed despite some of the changes. The big reshuffle proved it's possible. The precedent for rescheduling sports is there. Which is good, because we might need to use it.

Switching up the training and competition schedules in sport could preserve playing opportunities and player health without needing to build new facilities or buy all sorts of new equipment. In some cases, safe conditions for play can be found a week or month away from the current game days. If rescheduling is done early enough – that is to say, ahead of the competition season beginning and not as a last-minute measure – athletes and coaches will have time to adapt their training accordingly. Tickets can also be sold for a more climate-sure date, reducing the likelihood of cancellations or delays on the day.

This has worked in the past. The 2022 FIFA Men's World Cup in Qatar was hosted in November and December to escape the summer heat. Going even further back, the 1964 Tokyo Olympics were played in October because the summer was deemed too warm.* By shifting the schedules for these events way ahead of time, most of the foreseeable climate impacts were avoided altogether. And because everybody had enough forewarning, it didn't create as much disruption as in-tournament delays would have.

The difference between midday temperatures and evening temperatures are substantial enough to warrant same-day rescheduling. In 2019, World Athletics moved the marathon events at their World Championship event in Qatar to a midnight start time, to escape the wrath of the sun. Even so, both the men's and women's events saw several athletes drop out due to the punishing heat, and the finishing times were some of the slowest in world championship race history. Still, the outcomes would have been much worse had the race been run in daylight.

---

*This wisdom regarding summer temperatures in Tokyo was obviously lost on organizers of the 2020 Olympic Games, which saw the hottest games on record in July 2021 in the delayed Games after the pandemic.

Rescheduling doesn't necessarily only mean the date of the training and competition. It may include moving the timing of the games around, or extending the length of the broadcast so more in-game breaks can be offered to the athletes. If a two-hour football match is instead scheduled to take two and a half, then water breaks can be added to each half, for example. Purists may grumble, but these changes wouldn't meaningfully change the way the game is played. They would, however, ensure the athletes are kept safe. This is already happening in tennis, where new rules around water breaks are being trialed at tournaments on the WTA and ATP tours.

The biggest inconveniences of a reconsidered and more flexible schedule will come to those who travel to attend games. As Jessica Murfree explains, "Those disruptions to personal plans are what people hold on to. It's not so much the game being pushed back by a few hours or a day, it's everything else that now has to be rescheduled." If we focus on more local models of sport, where we prioritize local fans, and there's an understanding upfront from traveling fans that there could be changes, the sports sector can shift people's expectations around events. When the expectation is that things are flexible, people will adopt a more flexible approach. This may not happen overnight, but early research into the feasibility of rescheduling suggests it's possible.

Murfree adds that while there are so many logistics that make it challenging to delay a game, or relocate, or postpone, it is usually feasible. "That's because people in good faith will be willing to make accommodations. Most of the time, there's a lot of revenue that comes when the game is played, and that revenue is lost when it's canceled. Organizers are motivated to make this work for financial reasons, and fans are motivated to accept the changes to keep sport alive."

Emergency protocols

When rescheduling can't happen in plenty of time, then strong emergency protocols become necessary. For example, along the US Gulf Coast, which commonly experiences hurricanes, it is common practice for athletes and teams to relocate temporarily when a hurricane is on the horizon, so training and competitions are not indefinitely canceled. When Hurricane Ida hit Louisiana on August 29, 2021, the student athletes at Tulane University and the University of New Orleans were sent to train and compete at universities in nearby states. Similarly, the New Orleans Saints in the NFL left town and settled temporarily in Dallas, Texas. They played their opening game of the 2021 season against the Green Bay Packers in Jacksonville, Florida, while their hometown and home field recovered from the storm.

In an emergency, people need to move fast and decisions must be made quickly. So, after Hurricane Katrina shut down Tulane University for the whole fall season in 2005, the Athletics Department went through a multi-year process of designing and refining an emergency response plan. Next time, they'd be ready. It included everything from shutting down the sports facilities on their campus, to a contract with a university in Birmingham, Alabama that would house their staff and athletes for a prolonged stay. The emergency protocol included family accommodation for the staff and athletes. The team enlisted a Birmingham-based kennel and veterinarian to take care of pets. A list of grocery stores and restaurants near the Birmingham accommodation was drafted to make sure people knew where to find food. Moving a lot of people on short notice is not easy, but for Tulane's Associate Athletic Director, Chris Maitre, it was a case of following the plans when Hurricane Ida hit New Orleans on the 16th anniversary of Hurricane Katrina.

> The roadways are wild during the day, so we planned to stage everything during the day, gather everybody, then leave at night. We spent the day bringing in all the equipment from every sports field and facility, moving electric equipment off the floors and up above flood lines, and packing. Then, we lined up thirteen Mercedes Benz Sprinters, we hired nine coaches, we packed equipment into an eighteen-wheeler,

and we became a fully mobile department. We left at seven o'clock in a motorcade, arrived in Birmingham on Sunday morning. You can imagine how many stops we had to make, with so many people involved. We took up pets, we took up people, we took up family members of anybody who needed to be evacuated.

We had a young couple in our track team. And they had a tarantula. They had two fancy rats. I had no idea what those were. And they had a snake. I just told them, "okay, good luck with that". We put all of the pets in one vehicle. So that was a pretty smelly vehicle. But we took care of it.

We gave everybody gas cards, to make sure every vehicle could get to Birmingham easily. Those were already in my office ahead of time, we had them on hand just in case. It was a fairly well-orchestrated plan compared to what we had at Katrina. We never thought we would need it, but lessons were learned and we did.

Drafting an emergency protocol for the hazards you're facing is not an easy process, but it's worthwhile. Most insurance plans require you to have one, so there may already be emergency plans in place for some scenarios which can be expanded to include additional hazards and types of emergency.

The most important part of the emergency protocols, according to Maitre, is ensuring that everybody knows what their role is before the emergency happens. This requires the team, club, or organization to do a dry run – like a fire drill – to make sure everyone is clear on the plan. Step by step, review what needs to be done, how it will be done, and who is responsible. The same can be done with a bad-weather policy.

Inclement weather and policies

Most sports already have a rain policy. Lightning protocols are ubiquitous. These were designed to ensure that in poor conditions, officials don't need to think too much and can make quick decisions based on existing policies. It also manages expectations for athletes, coaches, and fans, and keeps everybody safe.

Weather policies can be expanded to include temperature and air quality. It sounds simple, and to some extent it is, but it's truly remarkable how rare these are. The Australian Open is one of the only major sports events to have a weather policy, because temperatures routinely float between 30 and 40 degrees Celsius (86 to 104 degrees Fahrenheit) during the January tournament. The policy was initially set up in the 1990s and called for games to be canceled if temperatures on the courts reached 40 degrees Celsius (104 degrees Fahrenheit) wet-bulb globe temperature (WBGT) (see page 45). In 2002, the policy was updated to stop play at 38 degrees Celsius (100.4 degrees Fahrenheit). In the last few years, additional changes have been made to the policy to include more breaks during the matches for athletes to access water and shade.

A similar policy is in place at Major League Soccer in the US, where heat breaks are implemented at the 30-minute and 75-minute mark if the WBGT readings are higher than 27.7 degrees Celsius (82 degrees Fahrenheit). In the same policy, officials are instructed to call a game delay if WBGT soars above 33.5 degrees Celsius (92.3 degrees Fahrenheit). Crafting such policies is not particularly complicated and sample policies have been made available from Sports Medicine Australia and the National Athletic Trainers' Association. For sports that involve animals, such as equestrian events, the animals' wellbeing must be considered as well.

As for air quality policies, a handful of organizations have adopted strict limits for maximum air quality – including Stanford University Athletics, which I mentioned earlier in the book – but for the most part, we're still seeing loose guidelines on air. This is partially because it's harder to identify a specific set of thresholds on various toxins, given the more long-term nature of pollution-related health impacts. More research is needed to figure out exactly how much is too much pollution, but at a minimum, we can start by setting policies that stop play as soon as the Air Quality Index (AQI) reaches what the World Health Organization considers "very unhealthy" or "hazardous".

Weather policies may seem like a simple adaptation item on paper, but implementing policies can be tricky and involve consultations with several interested parties: broadcasters, referees, coaches, parents, and athletes. It requires careful consideration of how potential delays and cancellations would impact all parties and guidance on how disruptions will be recovered.

And of course, it's hard to implement any policy without a clear way to gather information on the WBGT and air quality. Radars and weather awareness services are widely available, and that's a great place to start. Organizations with more resources can purchase their own wet-bulb globe thermometers, which sell for around £200 ($250) on the cheaper end, and air quality monitors, which range from £200 to £2,000 ($250–$2,500). This is an investment in the wellbeing of everybody on site, and the integrity of any weather policies.

Making a move

In some cases, staying in one place and adapting to climate change by adopting flexible schedules and weather policies, alongside emergency protocols, is simply not enough.

Relocating is not ideal for sports: place attachment is high. Most competitive sports are organized around cities and nations, with athletes representing their homes. Understandably, people aren't too happy about teams and clubs moving around. Ask my husband: he grew up a fan of the Montreal Expos baseball team until they moved to Washington, DC in 2004. He's still salty about it. Yet, with climate change eroding coastlines and delivering recurring disasters in some vulnerable regions, we could see more of that in future.

There's no denying that some places are just not suitable for some sports. For decades, athletes from developing countries with poor sports infrastructure have relocated to Europe, Australia, or North America to train in better facilities. This is how, for example, athletes from tropical nations have competed in the Winter Olympic Games: most had either moved at a young age and retained citizenship in the wintry country, while a few had been adopted into families from cold-weather countries and had lived there since early childhood. However, a whole new variety of athlete migration could begin with climate change. If the projection that 20 per cent of Olympic nations will have unplayable grounds by 2030 comes to pass, there could be a whole new slew of climate migrants joining the Refugee Olympic team or seeking transfers to other national teams.

Major sports events will also need to reconsider their locations. In November 2022, the International Olympic Committee announced it was postponing the decision on which place would host the 2030 Winter Olympic Games, so that its Future Host Commission could delve deeper into the developing science on climate change and warming winters. The Commission also wants to explore alternative hosting models. One option on the table is to rotate the Winter Games among a short list of four or five host regions. This would radically change the hosting costs by ensuring there are no "white elephants", and would allow the IOC to work with hosts that have suitable natural climates for winter sport. The downside? Restricting the Winter Games to just a few host regions could limit the growth potential of winter sports somewhat. Candidly, I love this idea. Sure, it has its downsides, but it locks in a future for winter sports and I count that as a win.

This is not the first time the IOC has made changes to the hosting model to accommodate climate change. On a call, Michelle Lemaitre, Head of Sustainability, and Jacqueline Barrett, Future Olympic Games Hosts Director, explained that already several steps have been taken to future-proof the Games. First, the IOC stated a preference for hosts with existing venues, which means not only that there will be no major construction efforts to create facilities that won't be used after the Games (this will be great for reducing the emissions and waste related to the event), but also that there are already people in the host region who know how to operate the facilities, which is ideal for the long-term development of sport in the region.

The other major change the IOC has adopted is to drop minimum venue sizes for Games hosting. This small edit is a big deal. In decades past, prospective hosts had to demonstrate that they could accommodate thousands of fans at every venue, and offer tens of thousands of available hotel rooms, because the Games were viewed primarily as a tourism enterprise. This led most hosts to build all sorts of new facilities: they may have already had a pool, but now they needed a big pool with loads of seating. Perhaps they already had a small stadium, but now they needed a very large stadium. With minimum sizes gone, a smaller town or city could feasibly host an Olympic Games with the venues they already have: no need for a new mega-stadium or 20 new buildings to form an Olympic village. Instead, they can use small sports venues – provided they have some spectator areas and media hosting capacities – while on the hospitality side, university dorms are on the table, existing hotels are in the mix, and Airbnb is the lead provider of accommodation. Considering the research that shows that previous host cities may be unable to host future Games due to climate change, the option to scale down the size of the event and make it fit into smaller cities will unlock a much longer list of prospective hosts and drive down the hosting costs. And for those watching from home, which is the vast majority of Olympic fans globally, the viewing experience will remain top-notch.

Sometimes the location changes in response to climate change are simple: move indoors. For example, in 2020, the Texas Rangers in the MLB opened a new facility costing $1.1 billion (£870,000) just 26 years after building their previous facility, Globe Life Park. The owners were quoted saying they'd done all they could to alleviate the effects of heat in the old open-air venue and concluded that for the game to be tenable in Arlington into the future, it needed to move indoors so the space could be air conditioned. While this move does eliminate some of the concerns around heat and precipitation, Jessica Murfree is quick to highlight the shortcomings of relocating sports to indoor locations: "Moving something indoors is just like putting a Band-Aid over it. And we know that moving something indoors is not climate-proof. Dirty air gets indoors, a flood will come indoors, and all of the people coming to your venue need to travel safely through the outdoors."

Still, relocating is an option, and while nostalgia is strong in sport, this option is attractive in cases where the current venues and locations are no longer desirable or become uninsurable.

Be prepared

It's impossible to overstate the importance of insurance in the face of climate change. If it weren't for insurance, several of the organizations mentioned in this book so far would have been crippled by the debts associated with building repairs, injury and medical payouts, event cancellations, and other damages. All sports organizations will have to check their insurance to ensure they're covered for hazards that are projected to arise in their area. In some cases, though, insurance alone won't cover it.

The Oakland A's have faced floods, storms, and wildfire smoke in recent years. President Dave Kaval has filed several insurance claims for damages associated with these hazards, but knows it's going to take more than that to keep the league fair and to keep his team competitive. He told me there have been "murmurs" among MLB owners of developing a league-wide contingency fund that all teams would pay into, and that all teams would have access to in the case of major hazards like hurricanes or fires impacting their season. "It would be a kind of self-insurance program to supplement each team's individual policies, without being specific to any one problem. The fund can't be earmarked specifically for flooding, or for electrical repairs, or for smoke damage: every team has its own set of climate hazards and this proposal would only work if all the owners got on board." Sharing the costs of damages would be a new form of cross-subsidization in competitive sport, which is already common in professional leagues as a way of keeping the game competitive. If the teams can share some of the profits, surely they can share some of the climate costs. The most recent expansion teams joining the MLB were the Tampa Bay Rays and the Arizona Diamondbacks in 1998 – so every team in the league has been involved since way before climate change started wreaking its particular form of havoc.

Sports organizations don't have to adapt alone. They can do it together, through cross-subsidization schemes like the one Dave Kaval proposed, and new league-wide or sport-wide policies. Sports can also lean on their insurers and their local authorities.

Work with the city
The good news is that most cities now have a climate adaptation plan. Those that don't, soon will. Increasingly, it's viewed as best practice in local governance to consider and adopt climate adaptation strategies that will reduce the risks for local residents and alleviate the costs of climate change, such as disaster recovery costs.

Collaborative opportunities exist for sports organizations to work with their cities to be part of broader adaptation plans. In an earlier chapter, I explained how this has worked in Puerto Rico and the US South, where sports venues have been earmarked to become shelters or emergency resource distribution sites during disasters. It's also possible the adaptation plans will change a sport's exposure to hazards. For instance, if a city is planning on eliminating traffic through the center of town, that could significantly improve air quality in that area and there may not be a need for a team or a stadium to have an air quality policy.* Or if a city is designing a new stormwater sewer system, your site's flood risk may go down. In short, it's best to stay on top of the city's plans. You never know when they might solve some problems for you or open doors to new solutions.

Take care of each other
Climate change is not just a physical phenomenon. It's also psychological. As individuals and groups experience climate change, either through anticipation of future hazards or through the trauma of previous ones, sport needs to make room for the full range of emotions that could come with it. You may have noticed that in most chapters, I brought up the mental health implications of the climate hazard. Living through climate change is traumatic and will bring more trauma as conditions worsen.

It's not possible to say exactly what those traumas and emotional challenges may be, or what types of support will be most helpful. However, it's never too early to add mental health services – such as mental health days for staff and participants, access to counselling services, and a list of local mental health resources – to the list of necessary adaptation steps. Sport and physical activity are also good for mental health, so offering incentives or time for employees and volunteers to use your facilities for free can be another great way to support mental health.

Already in my classroom, I'm noticing the impact of climate trauma. When I teach this content I can see that it weighs heavily on my students. This book may have the same effect. Our brains are not wired to take in copious amounts of negative information. Affording ourselves and others the space and time to process

---

*Unless, of course, the region also has ports, large manufacturing zones, or nearby forests that see wildfires. Those would all be good reasons to keep the air quality policy.

any feelings that come up will be the only way we can sustain conversations on this topic and stay active in climate adaptation over the long term.

Ultimately, the climate crisis will require everyone and everything to be more flexible. Easier said than done for a tradition-obsessed, highly structured sector like sport. Nonetheless, we must adapt to the hazards we are already experiencing. That's not all. We also have to act now to slow and stop climate change. Just because sport is beloved globally does not mean it gets a "pass" on climate action. Actually, it means the opposite: sport needs to be an early mover.

CHAPTER THIRTEEN

BACK STORY

Environmental efforts have been happening in and around sports for almost a century. To understand where we are now, it's useful to revisit where we've been.

Green Olympics

This story begins in 1929. It was early April in Lausanne, Switzerland, and the International Olympic Committee (IOC) was meeting to grant the hosting rights for the 1932 Winter Olympic Games. As the US was hosting the Summer Games that year, the country was also the frontrunner to host a Winter Games. Seven US candidate cities showed up to bid: Bear Mountain, Denver, Duluth, Lake Placid, Lake Tahoe, Minneapolis, and Yosemite Valley. For many small mountain towns, this would be their shot to generate the investment dollars needed to build new winter sport infrastructure and secure a strong tourism industry in a very economically fragile time – remember, this was 1929, the start of the Great Depression.

Dr. Godfrey Dewey, President of the very exclusive Lake Placid Club was the lead administrator of the Lake Placid bid. Knowing he needed his bid to stand out in a crowded pool of US candidates, Dewey made a series of lofty promises – chief among them, a new Olympic sliding track for bobsled. Franklin D. Roosevelt, the Governor of New York at the time (later to become President of the United States), wrote letters in support of the project, turning the small New York town into a frontrunner.

It all seemed promising, until the hosting rights were secured and things got messy.

Dewey's plan for the bobsled run was to be on the Lake Placid Club's site, within the protected Adirondack Forest Preserve. Immediately, the Association for the Protection of the Adirondacks (AfPA) filed a legal complaint with the New York State courts claiming that the project went against the "forever wild" clause of the state's constitution. A two-year legal battle ensued, with environmental activists routinely protesting during that period and making noise in the press. Eventually, in *Association for Protection of Adirondacks v Macdonald*, the Court held that a statute authorizing the construction of a bobsled run, requiring the destruction of 2,500 trees, for the 1932 Winter Olympics in Lake Placid, was unconstitutional. Dewey and the organizing committee were forced to find a location elsewhere.

That kind of pressure from environmental groups has never gone away. In nearly every Olympic host city, there has been a movement of what sport sociologist Dr. Jules Boykoff calls the NOlympic movement – organized groups of people who do not want these large, damaging events to come to town. In most cases, environmental activists are among the NOlympic crowd.

Fast-forward to 1970. Denver was awarded the hosting rights for the 1976 Winter Olympics. It was billed as an ideal opportunity to celebrate the US's bicentennial and Colorado's centennial anniversaries. But it took less than a year for the Games to be met with major dissent from local politicians. Months after the hosting rights were awarded to the city, State Representative Bob Jackson told the Associated Press, "We ought to say to the nation and the world, 'We're sorry, we are concerned about the environment. We made a mistake. Take the Games elsewhere.'" Dick Lamm, another State Representative who would go on to become Colorado's governor, told *Ski Magazine*, "Every time I ask a question about ecology, the Olympic people tell me, 'Don't worry, we are going to take care of that.' But a state which has never taken down as much as a single billboard to improve the environment is not going to run an Olympics which the ecologists would like." By 1972, the city withdrew from hosting and the Games were moved to Innsbruck, which had hosted in 1964 and had most of the facilities ready to go.

Germany also saw environmental groups put pressure on – and ultimately shut down – the Olympics over environmental concerns. In 1983, the mayor of Berchtesgaden and the local tourism director announced a bid for the 1992 Winter Olympics. Almost immediately, a local citizens initiative was organized against it. Dr. Amanda Shuman, a historian at the University of Freiburg, has been studying how this particular citizens' initiative was contextually different from those that preceded it in other countries. She and I work together through the Sport Ecology Group, so I called her to get the background story. She told me,

> The early 1980s were a unique time for the environmental movement in Germany. Acid rain was at the top of everyone's mind because Der Spiegel, the country's biggest magazine, decided to run a series of exposés on Waldsterben [forest death] with pretty aggressive headlines like "The Forest is Dying". At the same time, the newly formed Green Party rode that wave of public concern into their first seats in Parliament. Environmental groups watched this happen and were emboldened to act on different issues because there was more visibility and political support for their work.

Shuman argues that historically, anti-sport development efforts went against the political grain. But in 1970s Colorado, and again in 1980s Germany, there was some degree of political support behind anti-Olympics campaigns.

Pressure has come from other corners of the sports world, too. In South Korea, an intense period of golf course development in the 1980s led to a series of landslides and crop failures due to chemical contamination from nearby golf courses, and health issues for farming communities. Through the 1980s, farming villages began organizing to protest against the new golf courses, but it wasn't until the 1990s that the various small efforts converged into a movement, gaining significant media attention in the country. Mostly, these groups were unsuccessful and the golf developments went ahead as planned.

Despite repeated involvement and pressure by environmental groups to slow, move, or shut down sport development and big events through the twentieth century, it wasn't until the 1990s that sports organizations took up the mantle of environmental action themselves. In the same year that the word "sustainability" entered the global lexicon at the 1992 Earth Summit in Brazil, the IOC was facing challenges with the perceptions of the Olympic Games following the Albertville Winter Games, which were dubbed an "environmental catastrophe" in the local news, given the extraordinary distance between the different venues and the amount of traffic. The event was so spread out geographically that athletes and spectators were driving through the mountains from one town to the next, clogging up the roadways and emitting a ton of pollution in an otherwise quiet area of France. After that, the IOC knew it had to act to strengthen its reputation on environmental issues and align more closely with the growing global concern for a healthy environment. It's still not clear if they've succeeded on this, at least not yet – the Olympic movement is a big ship to turn, and these things take time.

The 1994 Winter Olympics in Lillehammer, Norway, are now viewed as the first attempt to create a "green" Olympic Games.* It was a tall order. After the environmental wreckage at Albertville, local activists in Lillehammer forced the organizing committee to adapt their hosting plans based on environmental concerns. The changes included a redesigned speed-skating rink that minimized impacts on a nearby bird sanctuary, a plan to prioritize the use

---

*In the 1990s, the Olympic hosting model shifted from hosting the Winter and Summer Games in the same year, to alternating years – so a major Games competition is held every other year. As a result, 1992 saw both the Barcelona Summer Games and the Albertville Winter Games, and then there was a quick turnover to the 1994 Lillehammer Winter Games.

of renewable building materials, energy-efficiency upgrades for facilities, and a recycling program at all venues.

The 1990s were a supercharged decade for sustainability across all sports, not just the Olympic Games. In 1993, the National Football League in the US launched their NFL Green campaign, which has seen every subsequent Super Bowl implement waste management and nature restoration projects. In 1994, the United Nations Environment Program (UNEP) created its Sports and Environment Program, to promote environmental awareness through sports and sustainable design principles in sports facilities and equipment manufacturing. Also in 1994, the Centennial Olympic Congress of Paris named the environment a "third pillar" of the Olympic charter, alongside sport and culture.

Later in the 1990s, the UNEP worked with the IOC to develop an "Agenda 21" for the Olympic Movement based on sustainability guidelines created by delegates at the 1992 Earth Summit in Rio. The IOC committed itself to promoting sustainability among its 206 member nations and 30 governing bodies for winter and summer sports, and to require sustainability plans from the hosts of its marquee events. This is only a commitment to "encourage" sustainability though, not to mandate it, as the IOC does not control operations among its members.

Despite these ambitions, the process of implementation has been a roller-coaster, with several sharp turns off-course. A 2021 study published by Martin Müller and colleagues at the University of Lausanne developed a model to evaluate the environmental sustainability of Olympic Games hosted between 1992 and 2020, and found that Salt Lake City in 2002 was the most sustainable, while more recent iterations at Sochi 2014 and Rio 2016 were the least sustainable. Part of the challenge for the IOC is that each host country is operating within its own sets of definitions, limitations, and government priorities, so sustainability often takes a back seat to tourism development and growth plans. Nonetheless, the findings of the study are interesting, as another sport sustainability movement was brewing.

Spreading the message

In the early 2000s, Dr. Allen Hershkowitz – once dubbed the Godfather of Green Sports by the Green Sports Blog – was working at the National Resource Defense Council (NRDC) as an industrial ecologist when he got a call out of the blue from Jeffrey and Christina Lurie, owners of the Philadelphia Eagles in the NFL. They had just finished developing Lincoln Financial Field, and they were calling because they wanted Hershkowitz's help to make it green.

Hershkowitz was a bit surprised but keen to help. He'd been a sports fan his whole life, so working with the Eagles seemed like a fun project. He started reviewing the Eagles' operations and supply chain. It didn't take long to identify the first problem: toilet paper. The Eagles were buying their bathroom tissue and some of their other paper products from paper mills that relied on forest that hosted eagle habitats. Eagles were being wiped out to produce toilet paper and other paper products at the Eagles' home stadium. Once the Luries learned about that, they immediately changed their paper products supplier. "The decision was instant," Hershkowitz recalls, "and that showed me they were serious about sustainability."

Around the same time, actor Robert Redford took an interest in greening sports. Recognizing that he had an opportunity to reach broad audiences, Redford contacted Hershkowitz and suggested that a whole program be created around sports, aimed at reaching audiences away from LA and New York and the big cities: this was the best opportunity to work with people in Kansas, Nebraska, Oklahoma, Texas, and the vast heartlands of the US. Since Hershkowitz was already doing this work with the Eagles, he joined Redford in reaching out to then-MLB commissioner Bud Selig. Hershkowitz recalls that "It turned out, Bud Selig was a spectacular environmentalist."

With Selig's support, Hershkowitz and his team at the NRDC worked to develop a league-wide sustainability program for two years, spending nearly a million dollars to produce 30 environmental guides for every baseball team in the league. Then, in 2007, the Major League Baseball Commissioner's Initiative on Sustainable Ballpark Operations was officially launched. It was the second league-wide effort to green sports, following the NFL Green program. The key difference between the efforts was that the MLB's program required all the franchises to act, whereas NFL Green at the time was restricted to league-hosted events such as Super Bowl and the Draft.

In 2009, Hershkowitz got another call out of the blue. This time, the call came from Jason Twill, who worked for Paul Allen, co-founder of Microsoft. Twill had heard about the NRDC's work with baseball and wanted to set up something similar for Allen's teams: the Seattle Seahawks (NFL), the Seattle Sounders (MLS), and the Portland Trail Blazers (NBA). Hershkowitz remembers smiling over the phone as he said, "You've got three teams in three different leagues. I'm working with the Seattle Mariners. That's a fourth team from a fourth league. Why don't we create a Pacific Northwest professional sports greening coalition?"

A few weeks later, Hershkowitz was on a flight to Seattle to meet with Twill in person. As they hashed out the possibilities for a regional sports alliance, they realized a national approach might make more sense: more teams, strength in numbers. The Green Sports Alliance was born in 2010 with six founding members: Seattle Seahawks, Portland Trail Blazers, Seattle Sounders FC, Seattle Mariners, Seattle Storm, and Vancouver Canucks.

Hershkowitz eventually left the NRDC to become the inaugural President of the Green Sports Alliance, whose members now number nearly 600 sports teams and venues from 15 sports leagues, mostly in North America. But that wasn't enough for Hershkowitz. He wanted to see this work go global.

So he started traveling to Europe, meeting with the International Olympic Committee, university professors, professional sports teams, and leagues. As time progressed, those trips became more frequent. Eventually, the trips resulted in Hershkowitz working with European colleagues to launch Sport and Sustainability International (SanSi), headquartered in Geneva, with a US office in Connecticut. Now, SanSi has members in 50 countries around the world.

Meanwhile, in the UK, a similar movement was underway. Dr. Russell Seymour had trained as a scientist in his early career, earning a PhD in biodiversity management studying giraffes in Africa. Coming out of his studies, he had hoped to find work as a conservation researcher, but when that didn't pan out, he took a job at Imperial College London teaching conservation science. One day, one of his students was recruiting volunteers to serve as ushers at Lord's Cricket Ground – one of the most famous cricket venues in the world – for an upcoming match. Seymour thought it would make for a nice day out, so he signed up. That was his entry into sport. And it stuck. Seymour ushered for six years, eventually moving to a full-time role at Marylebone Cricket Club (MCC), the club which owns Lord's.

One day, the CEO popped his head around the door of Seymour's office and said, "Do you really have a PhD? Come and work for me." Seymour became a special assistant to the CEO, working on a variety of projects, but there wasn't always a ton to do. After a while, he decided to fill his time by going back to his roots: examining the environmental footprint at MCC. He wrote up a report in 2006 on the organization's carbon footprint, recycling rates, water usage, and current efficiency efforts, and upon reading it, the CEO changed Seymour's title to Sustainability Manager, making it his full-time job. To the best of my knowledge (and Seymour's), this was the first role of its kind in the UK.

During Seymour's tenure at MCC, the organization moved to 100 per cent renewable energy, reduced single-use plastics, and overhauled the watering system to a state-of-the-art computer-controlled technology that included moisture readers in the soil and nighttime watering schedules to avoid evaporation in the heat of the day. He also started helping other organizations to get on board and do the same. In 2010, Seymour convened a group of 17 facility operators, sports clubs, governing bodies, and scientists to discuss how they could share information and tackle sustainability together. The meeting went well, and six months later another one was held – this time with 33 organizations participating. Slowly, the network grew, and by November of 2011, the British Association for Sustainability in Sport (BASIS) was officially launched.

"The discussions we were having were mostly operational back then," Seymour told me, "because that's where people were focusing their attention. What are the visible signs of sustainability? It's recycling and it's energy efficiency."

As the Green Sports Alliance, SanSi, and BASIS grew their memberships and programming on either side of the Atlantic, another trade association was forming on the other side of the world.

Dr. Sheila Nguyen was Co-Director of the Masters of Business (Sport Management) program at Deakin University in Australia where she researched corporate social responsibility in sport. In 2011 she traveled to the US for a short-term post as a visiting professor at University of Notre Dame. While there, she learned about the Green Sports Alliance and its work in professional and college sport. She knew something similar was doable back in Australia and New Zealand, so she called up her mentor, Malcolm Speed, the man responsible for popularizing professional basketball in Australia, and together they got to work organizing a symposium in 2013.

As it turned out, the timing of the symposium was good (or bad, depending on how you look at it). Australia was in the midst of a drought, and sports facilities across the country were struggling to keep their pitches green. Nguyen opened the symposium with the question, "How can we mitigate against climate change so that we can protect the places where we play for generations of play ahead?" Her punchy question raised eyebrows among the traditionally conservative business leaders in the room, but it worked. People were interested in being part of the discussion. Two years later, after more meetings and discussions, a group of foundational members signed on to form the Sport Environment Alliance. These members included high-profile organizations like Melbourne Cricket Ground, Netball Australia, Cricket

Australia, Australian Football League, Tennis Australia, and Golf Australia. Together, they worked on solutions to minimize sport's environmental impact and sought ways to galvanize the whole sports industry to prioritize the natural environment.

Hershkowitz, Seymour, and Nguyen have a lot in common. All are trained scientists with PhDs. All had worked in science or academic roles before entering the sports space. And they all saw incredible potential for sport to do more.

For the first 20 years of the green sports movement, from 1992 to 2012 or thereabouts, the focus was on operational improvements: reducing waste, switching to energy-efficient lighting, using less water, and measuring carbon footprints. These efforts were impactful. Consider how much toilet paper is used in a stadium with hundreds of toilets – it's a lot. Finding a toilet paper provider that uses recycled paper instead of fresh forests is a meaningful improvement. Or think about the water savings that can be achieved by implementing an irrigation system that cuts water use from 60,000 litres per night to 50,000. In one year, the facility will reduce its water consumption by more than 3.5 million litres. That's enough water to fill three Olympic-size swimming pools. But these efforts can be hard to communicate to fans and do little to leverage sport's sizeable platform to inspire their fans to act on climate change and build popular support for action.

No other industry captures the public imagination like sport. The 2019 Sports Around the World Report by Global Web Index (2019) surveyed 575,000 internet users globally and found that 83 per cent of people aged 16–64 watch at least one sport on TV. Some sports, like football (soccer) and cricket, enjoy fanbases in the billions. Even smaller sports, like badminton and rugby, have fan bases in the hundreds of millions. Recent studies show those fans overwhelmingly support climate action.

In 2020, the LifeTACKLE project surveyed European football fans and learned more than 90 per cent agree or strongly agree with the importance of (1) protecting the environment and natural resources, (2) preventing pollution, (3) respecting the Earth and living in harmony with other species, and (4) fighting climate change. Those are overwhelming numbers. A 2021 study by Global Web Index surveyed 9,763 sports fans aged in 15 regions and found that 69 per cent of Gen Z fans support environmental action and want to see sports stakeholders do more on this agenda. Sports organizations may worry that engaging in discussions around the environment is too political, and in some very specific regions, maybe it is. But on the whole, fans have given sports organizations a mandate to do more.

The United Nations Framework Convention on Climate Change (UNFCCC) has long known the power of sport to influence change. In 2016, just after the Obama administration dubbed October 6 "Green Sports Day" in the US, with the defending ice hockey champions Pittsburgh Penguins visiting the White House, conversations started at the UN to figure out how to educate the public about climate change. At the time, we didn't have a Greta Thunberg. There was no youth climate march. But there was sport.

Lindita Xhaferi-Salihu was working at the UNFCCC in 2016 when the organization was starting to think about how to move beyond the technical conversations and start to inspire the masses to pay attention to the climate crisis. The strategy at the time was to find those sectors closest to consumers, and work with them to pass the message on to the public. But first, they had to find sectors that had an interest in doing something about sustainability. So she started attending sport conferences, and one in particular was inspiring – the Sustainable Innovation in Sports Conference in Amsterdam, hosted by Claire Poole (we'll come back to Claire, keep her in mind). Xhaferi-Salihu was encouraged to see so many organizations represented there, but was also surprised to see how technical and specific the conversations were: it was all about paper, lighting, recycling, and operational goals. And the work was piecemeal, some out of the US, some out of the UK, but nothing truly global, nothing coordinated at scale.

A year later, Xhaferi-Salihu and her team decided to craft a framework that would offer the sports sector some guidance and ideas around how to tackle climate change and talk about it. The UN invited 40 sports organizations to Bonn and pitched the idea, but the response was clear: the sports leaders didn't want to talk about climate change if they didn't have their own house in order first. They didn't want to come across as hypocrites. They were happy to get involved once they had reduced their own footprint.

The Sports for Climate Action Framework, which came out of that meeting, was built with those two goals in mind: one was to walk the talk on climate change, and the other was to amplify climate messaging through sport. The first version of the framework released shortly after the Bonn meeting had targets and methodologies for reducing emissions, incorporated the Greenhouse Gas Protocol, and included a range of environmental considerations and action items. But sports organizations didn't like it. Xhaferi-Salihu recalls that "everybody said, no, we're not doing it. This is too complicated. We don't have people for this, we don't know what you're talking about."

So she went back to the drawing board and stripped it down. The Sports for Climate Action Framework 2.0, launched in late 2018 (and still in use today), offers signatories five principles to orient their environmental efforts. The principles invite sports organizations to:

- Undertake systematic efforts to promote greater environmental responsibility.
- Reduce overall climate impacts.
- Educate for climate action.
- Promote sustainable and responsible consumption.
- Advocate for climate action through communication.

This framework did the trick. It was easy, and vague enough for teams to feel comfortable committing to it, so they started to sign on, with the IOC, FIFA, World Sailing, Forest Green Rovers FC, the French Tennis Federation, Tokyo 2020 Olympics, and Paris 2024 Olympics as founding signatories. For a while, things really started to take off, with a slew of signatory announcements in 2019 and early 2020: the New York Yankees, the NBA, United World Wrestling, the Chicago Marathon, and Formula One. By the start of 2020, more than 100 organizations had signed on.

The pandemic proved a turning point for the green sports movement. Because of the break in the schedule, there was time to reconsider everything about how sports are played and delivered: how we schedule, how we host events, how we deliver the television viewing experience – all of it was up for renegotiation.

In the first months of the pandemic, my colleagues and I sat in Zoom meetings cringing over the seemingly inevitable return of single-use plastics at venues and people's reluctance to take public transit to games when fans returned to the stands. As it turns out, we didn't have much to worry about. Yes, single-use plastics came back with a vengeance (think of all those little sachets of condiments, lids and straws on cups, disposable masks that come in plastic, and multiply it all by 60,000 people at a sports venue). But the global public health crisis – combined with the Movement for Black Lives – gave sports organizations a new set of priorities. Something changed in 2020. Sports weren't just a form of entertainment anymore.

To my massive surprise (I wish I could say I saw it coming, but I was skeptical), sports started showing up in a big way to the climate movement. From my perch at the Sport Ecology Group, I watched it happen up close: sports organizations from all over the world started emailing our team, asking for the latest research and best practice guidance to supercharge their work. Our webinars went from having 20 attendees to 170 sports professionals in the middle of the pandemic. People in the sports sector were working from home, if they weren't furloughed, and finding ways to "build back better" seemed to be at the top of everyone's to-do list.

As the pandemic waned, and live sports returned in 2021, Xhaferi-Salihu announced a new set of targets for signatories of the Sports for Climate Action Framework. Recognizing the momentum that had built since the 2018 launch of the Framework, and that there was a need for a new North Star in terms of ambition, it was announced that signatories were now invited to join the UN Race to Zero – a pledge to halve emissions by 2030 and to reach net zero by 2040.

The Race to Zero targets were met with mixed reviews. Some sports organizations were excited: this was progress, it was something tangible to chase. Others were – and remain – hesitant, given how hard it will be to cut emissions by half and reach net zero in a sector defined by intercity and international travel. For elite and professional sport in particular, it will be tough to reduce emissions unless business is done differently. And then some people, like me, disliked it for a totally different reason. I didn't think it went far enough, so I tweeted that the initiative was "par for the course. It's good, and I'm happy orgs [sic] are committing to it, but it's not leadership … Countries with complex economies are committing to that goal. Do better."

This may seem harsh, but hear me out. While I'm happy Race to Zero is in place – any step is a good step – I'd love to see some of the big sports organizations best it.

Travel is a huge piece of the carbon puzzle for sport. Sports leagues at the professional and elite levels could make a decision to reorganize next season's schedule to reduce travel, and adopt policies that puts all inter-city travel on the ground unless it's longer than, say, five hours.* The sector has already accepted that a few hours of travel is acceptable in the air (think of the flight time from Boston to Los Angeles), so surely five hours on the comfortable, wide recliner seat of a private bus, or a first-class car on a train, would be fine, too.

---

*The French government has decided to pilot a flight ban between cities that can be reached by train in less than 2.5 hours. The ban was part of the country's 2021 Climate Law and was initiated by France's Citizens' Convention on Climate. It will remain in place for three years (to 2025), at which time, the government will reassess.

Switching to ground transport for short-distance travel could cut the team travel emissions linked to those games by upwards of 80 per cent. With this model, most of the Premier League games could use ground transport. Most of the Bundesliga, La Liga, Ligue 1, Serie A, too. And a significant percentage of the NBA, NFL, NHL, and MLB could also shift to ground travel for games. Is a bus as fancy as a jet? Admittedly, no. But the world is on fire. Decisions could be taken that cut emissions quite a lot, much more quickly than 2030 and 2040. Demonstrating that kind of forward thinking and commitment to sustainability would send a strong message to other sectors and to the massive fan base that it's not only possible, it's a priority.

With the Race to Zero set to begin, a separate conversation was being had over at the United Nations Environment Program – the UN division that had helped birth the green sports movement in the 1990s. Sam Barratt, the Chief of Advocacy and Education in UNEP's Ecosystems Division, was seeing the outcomes from the Sports for Climate Action Framework and trying to figure out how to activate the sports sector to tackle the nature and biodiversity loss crisis, too.* In late 2021, the UNEP and the IOC team approached me to help evaluate the prospect of a new framework dedicated to nature.

---

*The United Nations has several programs, funds, and conventions. The UN Framework Convention on Climate Change (UNFCCC) is one of them, and the United Nations Environment Program (UNEP) is another. When an initiative is launched by one unit, it will typically be specific to what that unit is responsible for, because that's within their remit. The UN recognizes a triple planetary crisis of climate change; nature and biodiversity loss; and pollution as being equally important areas to address, but different units are responsible for each. They work together, but these are big offices with loads of people, working from headquarters on different continents (UNFCCC is based in Bonn, while UNEP is based in Nairobi). The UNFCCC had set up the Sports for Climate Action Framework to address climate change. But as of 2021, when Sam Barratt started thinking about this, no framework existed for addressing biodiversity loss or pollution.

For four months, my research team conducted focus groups in English, French, and Spanish with representatives from more than 100 sports organizations around the world to ask about their sustainability work. We learned that only a handful of them had a staff person or committee dedicated to environmental efforts and a meagre 32 had formal environmental sustainability plans or strategies. We also found that most sports organizations had little capacity to take on more work, regardless how worthy the cause. However, participants agreed that nature is important and worth protecting, and they're seeing nature changing before their eyes – especially in the mountain and water sports. The big takeaway for our team was that there's an appetite for a nature agenda but it has to be easy.

The Sports for Nature Report, which emanated from that large focus group study, was first launched in November 2022 at Lord's Cricket Ground and again later that month at the IOC headquarters in Lausanne. Weeks later, the International Union for the Conservation of Nature took the baton and launched the Sports for Nature Framework with four principles of its own:

1 Protect nature and avoid damage to natural habitats and species;
2 Restore and regenerate nature wherever possible;
3 Understand and reduce risks to nature in your supply chains; and
4 Educate and inspire positive action for nature across and beyond sport.

It's purposefully simple – exactly what the doctor ordered.

As I write this, in July 2023, just under 300 sports organizations have signed on to the Sports for Climate Action Framework, including some heavy hitters, like the NBA, World Tennis Association (WTA), IOC, FIFA, and the list goes on. The new Sports for Nature Framework is also gaining momentum, with some 30+ signatories as of Fall 2023.

While it may sound like an impressive number, 300 represents only a very small fraction of all sports – think of professional football in Europe, for example. The European Leagues website proudly announces that it covers 40 professional football leagues with more than 1,000 clubs. Or consider that the National Collegiate Athletic Association in the US has 363 university athletic departments competing in its Division 1, with another 750 schools competing in the lower divisions. Imagine the number of youth sports clubs, sports events and tournaments, and independent sports venues in the world. In context, 300 seems like a painfully low uptake.

There are a number of reasons why sports organizations don't want to engage in environmental sustainability. Some don't feel they have the time or capacity to take on more work, which is perfectly valid in a sector where most of the money is concentrated at the top echelons while sports at lower levels are left to make do with a limited budget and skeleton staff – if there's any paid staff at all. Plenty of sports organizations, especially at the youth level, are volunteer-led. Others view sustainability as too science-y and complicated, and worry they don't have enough technical ability to get involved. This is fair enough, when you consider that most people working in sports were never trained on sustainability practices; it's not a core part of coaches' training anywhere in the world – as far as I know. So yes, we're asking sports managers and coaches to think about something they've never been tasked with before. And that will come with a learning curve. I expect this process to take a while.

Overall, though, the green sports movement is decidedly on the right track. Reflecting on the progress to date, Hershkowitz said in a podcast interview in 2021, "I think, actually, over the last ten years, the sport and sustainability movement has been one of the most effective sectors in the environmental advocacy world, especially in North America, where our government has been outright hostile to environmental progress."

From where I'm standing, it's clear a lot more has been happening in recent years. The sector is moving forward, and moving together. Now we have to pick up the pace.

GREEN SPORTS

Lew Blaustein has been writing about the green sports movement on his blog since 2013, offering a play-by-play analysis of sustainability efforts in sports. Mainly, his focus is on American sports, but he dabbles in international stories, too. In late 2019, as the Green Sports Alliance was approaching its ten-year anniversary and Blaustein was pondering the work to date, he met me for tea in a café near Central Park on NYC Marathon weekend. We spent two hours sitting in that café, hashing out where sport has been and what comes next in the climate movement. He suggested to me that we may have stepped beyond Green Sports 1.0, which he describes as "cleaning up our own house", and that perhaps we were finally entering the era of Green Sports 2.0: "using our voice".

I like to make this even simpler: sport has to reduce its footprint and increase its brainprint – the amount of attention we draw to climate change and other global issues.

Footprint first

Reducing sport's environmental footprint has mostly happened by lowering energy and water use, and diverting waste from landfills. While we know this work is happening in sports facilities across North America and Europe, it's hard to know who is doing what. In 2019, 86 per cent of Standard and Poors (S&P) 500 companies published corporate sustainability reports, up from 20 per cent in 2011. That same year, my colleagues in the Sport Ecology Group, Dr. Brian McCullough, Dr. Sylvia Trendafilova, and Dr. Jamee Pelcher reviewed the communications, websites, and publications of professional sports franchises in the US and found that only one had published a sustainability report. An additional 42 organizations highlighted their environmental initiatives on a dedicated page of their websites. The sector was moving – but slowly and quietly.

Still, a few examples of extraordinary waste, water, and energy management stand out. For more than eight years, the Waste Management Phoenix Open (yes, it's literally sponsored by a waste management firm) has been dubbed the largest zero-waste sports event in the world.* In partnership with the Thunderbirds Charities, tournament vendors and the PGA TOUR, Waste Management leads operational efforts to divert 100 per cent of waste from landfill through recycling, composting, donation and energy conversion. In 2022, 63.5 per cent of waste was recycled, 24.5 per cent was composted, 10.3 per cent was incinerated (waste-to-energy), and 1.7 per cent was donated. Of the donated goods, 4,060 pounds (1,800kg) of food was sent to local non-profits, 5.6 tonnes of building materials were sent to local construction and repair organizations, and more than 90,000 square feet (8,000 square meters) of signage was repurposed. In terms of water conservation, 50 million gallons (227 million litres) of water were restored thanks to water sponsors. The event was also run on renewable energy.

A similar program is underway at the annual TCS New York City Marathon event, which was able to divert 75 per cent of the 339 tonnes of waste the organizers collected at the 2019 event. I was at that race with a group of 100 students from SUNY

---

*Zero waste does not actually mean zero waste in the sustainability context. It actually means 90 per cent or more of waste was diverted from the landfill, meaning it was reused, recycled, donated, or incinerated.

Cortland, and can attest to how much waste there was and how impressive it is that most was diverted from landfill. It may sound like "just recycling" but I can't even pull that off in my own kitchen (despite my best efforts), let alone for an event that runs through five boroughs with 50,000 participants and hundreds of thousands of fans along the route. At the end of race day, as at all NYRR (New York Road Runners) races, leftover pre-race, course, and post-finish food and beverage items are collected and donated to City Harvest, New York City's first and largest food rescue organization. My students and I were sent to collect food from different areas of the finish line and carry it to back to a City Harvest truck to be sent to the 1.2 million food-insecure New Yorkers the organization serves.

On the venues side, Mercedes Benz Stadium in Atlanta set a standard as the first sports stadium to achieve LEED Platinum certification in 2017 for their 4,000 solar panels, electric vehicle charging stations, water recovery cisterns, energy-saving design, and waste management program. Two years later, Climate Pledge Arena in Seattle set a new first with their name: it's the first venue named after a campaign rather than a company, place, or person.

Climate Pledge Arena became the first International Living Future Institute zero carbon arena in the world to commit to no fossil fuel consumption in the arena, solar panels on the atrium and parking garage, and public transport partnerships. The building is also the first in the NHL to commit to eliminating single-use plastics by 2024 (there will be so many bamboo forks and knives in that building!) and for harvesting Seattle's rain in a 15,000-gallon (68,000-litre) cistern to feed the world's "greenest ice". At every turn, they're talking about it: the venue releases reports and press releases for each new improvement, and runs events and educational programs for others in the sector to learn and get up to speed.

Climate Pledge Arena is owned by Oak View Group, a lively young venue management company that is setting new standards across every area of sports venue sustainability with their Green

Operations and Advanced Leadership (GOAL) program, which provides venues with the tools and resources needed to accelerate environmental goals. Oak View Group's CEO Tim Leiweke is passionate about this vision. On stage at a *Sport Business Journal* event in New York City, Leiweke told the crowd, "We've got ten years, and if we don't fix it in ten years, this is over and there's no turning this back. In ten years, we are not going to be able to say we made a mistake, let's go fix it. We cannot fix the Earth in ten years if we don't do it today." This sense of urgency is exactly what's needed to drive change.

In Europe, the Sport Positive League Tables run by Claire Poole have kept tabs on environmental efforts by professional football clubs since 2020. Each year so far, Tottenham Hotspur have won their Premier League table (or tied for first) for their efforts to get fans to use public transport on match days, as well as embracing clean and renewable energy and reducing single-use plastics, among other initiatives. Together with Sky Sports, the club hosted "Game Zero" in 2021, billed as the "world's first net zero carbon football game". In a post-match report, the club announced that Game Zero goals were achieved by transporting players on biodiesel coaches, encouraging fans to walk* or use public transport, powering the event with renewable energy, serving local and sustainably sourced foods with 94 per cent more vegetarian meals compared to an average game, and reducing emissions linked to the broadcast by 70 per cent.

In the Bundesliga, VFL Wolfsburg tops the table, while Olympique Lyonnais dominates the sustainability game in France's Ligue 1. The Sport Positive League Tables assess clubs on their use of clean energy, water and energy efficiencies, transport, single-use plastics, waste management, plant-based food offerings, biodiversity

---

*The post-game reports calculated that fans walked a combined 36,000 miles to the game – which brings the song "500 miles" by The Proclaimers to mind. New "Game Zero" theme song?

work, and engagement efforts with fans and the public. What I find most inspiring in these tables is that no team in any nation's top league scored zero. Meaning every team was doing something. And something is better than nothing.

One particularly cool new development in sport's circularity journey is the innovative carbon fiber recycling project led by the World Sailing Trust. Carbon fiber is ubiquitous in sports: it's in rackets, skis, boats, bicycles, bows, bats, hockey sticks, and golf clubs, to name a few. But this is a material that cannot be recycled. When your carbon fiber equipment breaks, most of that gear ends up in landfills. Recently, a group of sports federations and sports brands – including Wilson Sporting Goods, SCOTT Sports, Starboard, and OneWay – partnered with Lineat Composites and the University of Bristol technical research center to find a way to recycle carbon fiber. So far, the project has managed to pioneer and demonstrate a technique to reclaim broken or failed carbon components from one piece of sporting equipment and turn it into uni-directional fiber tapes. These new technical carbon tapes will then be supplied to component manufacturers within the alliance to be integrated into new components. A typical example would take a broken carbon fiber bike component and use the fibers to make new tapes that would find a second life in a ski pole or a tennis racket.

Carbon emissions

Two decades in, and Green Sports 1.0 isn't over, it's ongoing. Solving operational issues to reduce sport's environmental footprint was and remains at the top of most sports organizations' to-do lists because they know they'll get scolded if they speak out on climate change without having their own house in order. And while some of the operational stuff is easy to see and understand, because the venue staff can see the piles of trash and pay the utility bills, the carbon emissions point is still a tricky one.

With carbon emissions, there are two tasks. The obvious one is to reduce emissions. For members of the Sports for Climate Action Framework, the marching orders are clear: reduce by 50 per cent by 2030, and reach carbon neutrality by 2040. If sports want to achieve this, it'll take more than energy efficiencies in the buildings, waste management projects, and switching to a renewable energy supplier. To drop emissions by that much, sports is going to have to tackle its travel problem. I brought this up already in the last chapter, but travel really is the biggest piece of the puzzle for sport's decarbonization journey, so I'm going to dive deeper.

In a 2021 study, Seth Wynes at Concordia University calculated that in the case of North American professional men's sports (for example, football, ice hockey, basketball, and baseball), emission reductions of up to 22 per cent could be achieved if they adopt some of the same travel policies they employed during the COVID pandemic. For instance, if leagues cancel overseas games, arrange the schedules by region, and increase the number of consecutive games between teams, that will lead to big wins in emission reductions. The other option is to reduce the length of the season, which has the added benefit of opening up more time for rescheduling in the case of inclement weather.

Fan travel is another sticky one. At mega-events like the Olympics and the FIFA World Cup, fan travel accounts for 80 per cent or more of emissions, largely because these events attract so many international fans who fly. Focusing events on local crowds and reducing the number of tickets available to out-of-towners could drop emissions by significant amounts, pretty quickly. It doesn't work quite the same way in more minor sports leagues, because the majority of fans at your average league game are local. Still, finding solutions to the fan travel conundrum will be a key piece of this puzzle, no matter how you look at it.

Some of the most exciting work is coming out of the Paris 2024 Olympic Committee, which promises a carbon-positive

Games.* The French capital has banned non-essential through-traffic from its city center effective in 2024, making 5.4 square miles of the city straddling both sides of the Seine much greener and cleaner. They're also adding bike lanes and bus routes, and 95 per cent of the venues will be existing facilities or temporary builds, so only two new builds are needed. Honestly, we haven't seen anything better than this. Still, there will be loads of tourists (it's Paris, there are always tourists, with or without the Games). So the organizers have committed to offsetting all remaining emissions.

The idea of carbon offsets is simple: an individual or organization can balance out their carbon emissions by investing in projects that reduce greenhouse gas (GHG) emissions or remove carbon dioxide from the atmosphere. Essentially, offsets are a way of paying to reduce the carbon footprint of one's activities. In theory, it makes sense and unlocks much-needed finance for nature-based solutions to climate change and drawdown projects that otherwise wouldn't be particularly sexy investments on the private market. The projects range from tree planting to coastal restoration, from green energy infrastructure to carbon capture and storage technologies. There are some great, innovative projects on the carbon market with important co-benefits: some create jobs for locals, clean the local water, ensure food security for a community, preserve biodiversity, and the list goes on.

However, the carbon market is poorly regulated (where there are any regulations at all) and so there are good and bad offsetting projects. It can be tricky to navigate the sea of offerings to find

---

*I have been outspoken in recent years about how I don't think it's possible to have a carbon-positive Olympic Games and that this language is misleading and potentially detrimental to the broader movement. It's great to see the ambition to be very low-impact, but 'carbon-positive' is just not realistic in the context of an international sporting event with hundreds of thousands of tourists and participants.

projects that are legitimate and do what they say. In some cases, investigations by legal firms and journalists have revealed that projects are overstating their benefits. Others have been caught double-selling, meaning the owners of the project sell one buyer a certain number of credits (let's say it's 10 credits for 10 tonnes of carbon, assuming the standard 1 credit to 1 tonne ratio) and then sells that same 10-credit allotment to another buyer. Without very strict and routine accounting, it is tricky to track down those kinds of nefarious sales.

The other problem with offsets is that some view them as a "get out of jail free" card. Instead of changing business practices, organizations buy offsets and call it a day. Reductions have to come first. And in the Olympic Movement, according to Michelle Lemaitre at the IOC, they do. But that leaves a large grey area on what is considered "residual emissions", which is to say what could be reduced but would be a hit to the economic impact of the event, or the social benefits, or the experience, or the sponsor revenues, and so it doesn't get cut.

Mark Cuban, owner of the Dallas Mavericks in the NBA, has been quietly offsetting his team's travel for several years. I asked him about it over email, and he confirmed the main goal is to "try to reduce our carbon footprint like everyone else", but he added, "I buy carbon offsets to offset our usage and travel. It's not perfect. But it's a start." It certainly is.

Offsets are far from perfect. Unfortunately, in sport, we're stuck with them until the airlines come up with electric planes or we change the event structures to not require as much travel. There are several initiatives the sports sector can do on its own, but changing the travel and shipping industries is not one of them. For now, offsets are fine. It's a 6 out of 10, better than medium.

Making new roles

In 2023, my colleagues and I started looking around the sector to figure out what else can be done in-house. What are the next steps? To answer that, we began collecting data on how many sustainability professionals are working in sport, what their day-to-day work looks like, and which departments they're working in. After all, without the staff to do this work, it just won't get done.

I'm friends with several people in these roles, and can attest they are moving mountains. Still, the to-do list is endlessly long, and with small teams and limited resources, there is only so much that can be achieved. Every sports organization, at the professional and elite level, has teams of employees working in operations, facilities, marketing, sales. Sustainability usually has one person, if that. We need more.

It's hard to recruit new people into these roles, as most job descriptions are asking for candidates with experience doing environmental impact analyses, or carbon accounting, or waste management expertise. The expectation is that candidates should have management or engineering degrees. The challenge is finding these unicorn workers who have the right experience, because nothing quite prepares a person for working in sport, where not only does everything move fast, it happens in a fishbowl. The world watches and criticizes sports organizations as a pastime.

Marine biologist Dr. Ayana Elizabeth Johnson has convincingly argued that every job can be a climate job, if we want it to be. She even created a handy Venn diagram to help individuals find their opportunity to contribute to the climate movement. It has three overlapping circles: 1) what brings you joy? 2) what work needs doing? and 3) what are you good at? So this is me telling you, dear reader, that if sport brings you joy and you're good at project management and environmental work, this sector has *a lot* of work that needs doing.

Recently, a slew of new trainings and university classes have popped up to train the next generation of climate workers in sport. I've been fortunate to be involved in some. In 2021, Joanna Leigh, Olympic gold-medalist in field hockey at Rio 2016 and carbon management professional, was working as the carbon manager for the Birmingham 2022 Commonwealth

Games. Over several calls and walks through Hampstead Heath in London, we discussed how best to upskill the next generation in this space. She proposed creating an eight-hour professional development training course for The Carbon Literacy Project, a Manchester-based non-profit on a mission to teach as many people as possible about how carbon works, so everybody can do a little bit more to reduce their footprint. Over the course of a year, we (mostly she) created the toolkit and piloted it with learners from six countries. It was formally launched in April 2023, and is freely available to anybody who wants to learn more.

I've also been involved in creating the world's first Master's program in Sustainable Sport Business at Loughborough University London. As I write this, our pilot year is underway, and the registration numbers are climbing for next year. The program takes students through modules on sustainable innovation, traditional and emerging business models, leadership for change, sport ecology, and research methods, to name a few. At the end of the modules, students write a dissertation in collaboration with an industry partner. I'm hopeful programs like this will prepare a new group of professionals for the climate-changed future of sport.

The role of sponsorship

It's not enough to have the right people and some understanding that climate action is "the right thing to do". Sport needs more motive to change its ways. In some cases, the pressure is coming from legislation. In others, it's coming straight from the business partners.

In recent years, sports sponsors have begun demanding more from the properties they support. Aileen McManamon, Chair of the Board at Green Sports Alliance and Managing Partner at 5T Sports, has been tracking this trend for over a decade. As she explains it, "Every major global corporation faces reporting pressures on social and climate issues – whether legislated or de facto from investors, employees and consumers. These corporations have invested hundreds of millions of dollars into their brand. They simply can't afford to 'trust' their sports partners are doing the right thing."

So the sponsors have started asking questions about the sports organizations' efforts. McManamon says this type of once-gentle pressure on sports organizations is getting more intense and more frequent. "In one case, a partnerships sales professional shared that since 2020, social responsibility questions are being raised in every conversation with existing and prospective partners."

Moving forward, McManamon expects sponsors will make social and environmental sustainability performance part of their contractual agreements with sports properties, especially considering that some of the best performing sponsorships also promote the same values. Values-based partnerships, which are intended to be activated around societal issues, are trending toward higher valuations, two to four times higher than your typical deals.

This trend is attracting new entrants to the sports sponsorship market that previously weren't investing in sports, like socially conscious Aspiration Bank, who signed a $300 million (£240 million) arena-naming sponsorship deal with the LA Clippers in 2021, and Longi, a solar and hydrogen energy company who signed a multi-year deal with the Association of Tennis Professionals (ATP) in 2023, becoming their official solar and hydrogen energy sponsor – a new category. McManamon views this as a strong approach to sponsorship: "These new 'breeds' of partner are focused on long-term objectives. As a result, the deals are more likely to be stable and hold up for the deal term; more so than some of the more fleeting partner categories of cryptocurrency and online wagering." The shifting sponsorship market is putting pressure on sports organizations to up their game on social and environmental work, and might be exactly the push the sector needs to move faster.

In one stand-out case, the pressure to perform on environmental matters is coming from the league itself. SailGP, a new competitive sailing series, has a unique system for managing environmental work: they make it part of the competition. Athletes compete on the water and in their communities, racing to make the biggest positive impact.

It's called the Impact League, and teams earn points for creating a sustainability strategy, adopting clean energy and sustainable technologies, reducing fuel use, choosing sustainable suppliers for their merchandise, reducing waste, choosing low-impact hotels and accommodations, adopting a plant-based diet, and engaging fans on environmental and social issues.

Isabella Bertold, who sails for Team Canada in SailGP, admits doing the sustainability work "sometimes feels like a chore and is not always easy, but we do it because we're incentivized to. We're motivated by the competition." The team at the top of the leaderboard after each SailGP competition earns a "golden ticket" to switch their crane time at the next event – the time that their boat is slated to get on or off the water. When Team Canada reached the top of the leaderboard in the 2022 season, Bertold remembers using the golden ticket to get off the water about an hour earlier than they were supposed to one night. "That was an hour more sleep between race days. It was worth it. We performed way better in the next day's race because of it." At the end of the season, the points from each team's full season of impact work are tallied and the winner of the "Podium for the Planet" receives a donation for their non-profit or charity partner.

Green Sports 2.0

SailGP's Impact League bridges operational sustainability improvements and the communications side of things. This is where Green Sports 2.0 begins.

Green Sports 2.0 has a wholly different focus: instead of zeroing in on the operations of sport, it fixes on the massive sports platform. In 2020 and 2021, I watched from the sidelines as a surge of sustainability reports, green sports campaigns, and media stories gained visibility in the busy sports space.

Teams started fan engagement campaigns to encourage fans to take public transit, reduce waste, and participate in the team's sustainability work. It sounds simple, but it took a long time for teams to start talking about sustainability. Many worried it would put their fans off. My co-director at the Sport Ecology Group, Dr. Brian McCullough of Texas A&M University, has been following these campaigns and measuring their impact. He's got good news for sports marketing departments: these campaigns work.

"By comparing pre- and post-intervention surveys of fans, we have demonstrated that fan engagement campaigns or interventions work. The campaigns do influence fans to behave more sustainably at sporting events, and in other studies, these behaviors transferred to their everyday lives." There's just one caveat. The initiative or "ask" has to make sense to fans. Perhaps it's a water-saving campaign in a water-stressed area, or a recycling campaign with beer cans being run at the bar. If the fit is not there, and the team has to explain why the initiative makes sense, it won't work.

Even better news: campaigns that encourage fans to engage in environmental initiatives can help teams pick up new fans. McCullough's work has shown that lower-identified fans – the people who will go to a game with their partner or a group of friends, but who could just as easily have decided to go to a movie that night – were likelier to go to more matches, listen to more broadcasts, and buy more merchandise if the team promoted sustainability values that aligned with their own. With awareness and concern for climate change skyrocketing among the younger generations in every part of the world, running environmental campaigns is a safe bet in sports right now.

The World Wildlife Fund (WWF) knows this. For the last three years, on World Wildlife Day (March 3), WWF teamed up with some of the world's best-known companies, NGOs, and sports teams by inviting them to remove images of nature from their branding, in a bold effort to highlight the emptiness of a world without nature. Using the #WorldWithoutNature hashtag, the campaign highlighted the dramatic loss of biodiversity globally. In 2022, 330 brands took part in the campaign, among them nearly 50 sports brands with animal mascots. The campaign reached more than 100 million users on Twitter. Sports brands were especially high-performing in the campaign, delivering 16 of the top 20 tweets in terms of engagement.

In some cases, it's the fans that are leveraging the sports platform to get the word out about climate change and sustainable options. Pledgeball is a non-profit based in the UK that is galvanizing football fans around sustainable behaviors. Users register a profile on the Pledgeball website, identify the team they support, and make a pledge to reduce emissions by switching to recycled toilet paper, taking public transit, shortening their showers, or one of the dozens of actions listed on the site. Each weekend, the Pledgeball staff tally the amount of carbon saved by the fans of each team, and announce the winners after each match on the football calendar. So far, the Pledgeball team has calculated emission reductions of more than 20,000 tonnes of carbon dioxide equivalent. Fans are a powerful force, once tapped.

Part of Green Sports 2.0 is getting the news media involved. Adam Silver, Commissioner of the NBA, once said that 95 per cent of sports fans won't ever go to a game or tournament because it's just too expensive or they don't live near a team. So they watch at home. Proponents of sustainability and climate action frequently cite the sports sector's size and platform as an opportunity to leverage sport for public engagement on environmental issues. However, the potential for sport to influence fans on issues of climate change is, for many, limited by the lack of media coverage of efforts.

Given the potential of the media to educate the public on climate change and the unique platform of sports to capture the imagination and inspire fans to adopt environmentally friendly behaviors, the intersection of mass media, climate action, and sport is a particularly dynamic and high-stakes arena for communicating environmental issues to the public. From 2017 to 2020, I systematically collected news media articles about sport and environmental sustainability published in English print media outlets (newspapers and magazines). In those three years, coverage grew from 11 articles in 2017 to 41 articles in 2018, 237 articles in 2019, and 240 articles in just the first quarter of 2020.

Unfortunately, in those same three years, there were several missed opportunities to tell climate stories through sport or to amplify sustainability work in sport. Even when there was an opportunity, potentially, to talk about climate change, because there had been a storm or because there had been extreme heat, sports journalists were not making those connections.

Things have changed since 2020. It seems as though suddenly, major media outlets are mainstreaming the climate conversation in sports. In 2022, *Sports Illustrated* announced a new climate desk, Sports Illustrated Climate. The coverage has been sparse – only three articles in the first three months – but still a step in the right direction.* *The Athletic* has climate stories in its primary coverage; at ESPN, Dan Murphy has posted several stories about climate impacts; BBC Sport has hired Dave Lockwood as its first Head of Editorial Sustainability; and Sky Sports has a vocal supporter of sport sustainability work in presenter David Garrido. I've personally been in touch with journalists working for media outlets on every continent to provide background information and interviews about this topic. Slowly but surely, coverage is growing and reaching the millions of fans we know are there.

It may be too soon to say what Green Sports 3.0 will bring, but if I had to wager an educated guess, I'd say it has something to do with influencing change among the vast sports supply chain. After all, sport doesn't just influence fans and command media attention. It also serves as the most important marketing mechanism for a huge swath of brands from some of the world's dirtiest industries: oil, aviation, fast fashion, and fast food, to name a few.

---

*It's also worth noting that *Sports Illustrated*, like many other major news outlets, faced severe financial challenges in late 2022 and early 2023, laying off several staff.

CHAPTER FIFTEEN

IN BED WITH BIG OIL

When I moved to the UK during the pandemic, the talk of the town in the sports world was the trend of Saudi, Emirati, and Qatari ownership proliferating in football (association football, aka soccer). Newcastle United was purchased by the Saudis. This threw a spotlight on Manchester City, which is owned by Sheikh Mansoor of the Emirates. As I write this, Manchester United are weighing bids from Qatar's Sheikh Jassim Bin Hamad Al-Thani and Sir Jim Radcliffe, British billionnaire and CEO of INEOS – a "global chemical company" – whose products include chemicals, polymers (read, plastics), oil and gas. These developments troubled fan bases, who are increasingly aware of – and concerned about – climate change. Islamophobia and racism also played a role here. Each takeover made the news for several days on end, stirring controversy.

In the same period of time, questions were being raised about the FIFA World Cup of 2022 being hosted in a petrostate. This, despite FIFA being "committed" to sustainability since the early 2000s. Talk about mixed messages.

As one of the resident sustainability experts in the sports world, I started getting questions from the media about the tensions between Big Oil owning and sponsoring sports, and the sustainability pledges and announcements being made by these same professional sports franchises and mega-events.

And my answer wasn't simple, because this is messy. But I'll start with this: oil ownership in sports (and sponsorship, but we'll come back to that) isn't new. Not even close. And some of the most egregious examples of oil money in the sports scene come from my own backyard. So I'll start there.

Big Oil takes over

The National Football League (NFL) is a 100-year-old organization that owns a day of the week (and is slowly creeping into the other days as well). Every week, 16.5 million viewers tune in to watch their favorite sports teams go head to head (literally and figuratively). And you can't mention the NFL without mentioning the Super Bowl, the biggest annual sporting spectacle in North America, wherein the best of each conference meet on the gridiron to win the coveted title of world champions – despite the fact that the sport isn't enjoyed anywhere outside the US.

While the NFL was founded in 1920, the first Super Bowl wasn't played until 1967, between the Green Bay Packers and the Kansas City Chiefs. What most people don't know is that the Super Bowl and the NFL as we know it wouldn't exist today if it weren't for the voracious greed of a few oil tycoons. Let's stick with the Kansas City team here for a moment.

In 1959, Lamar Hunt, son of oil tycoon H.L. Hunt, and his friend Bud Adams, were interested in bringing an expansion team to the NFL, but were unsuccessful. Pivoting, Adams decided to pitch a few other wealthy men on starting a league of their own. The "Foolish Club", as they were called by local newspapers, founded the American Football League (AFL) in 1960. Lamar Hunt founded the Dallas Texans, which later moved and became the Kansas City Chiefs, while Adams founded the Houston Oilers, which later became the Tennessee Titans. The other six members were Harry Wiseman, who founded the New York Titans, now known as the New York Jets, Bob Howsam with the Denver Broncos, Barron Hilton with the Los Angeles Chargers, Ralph Wilson Jr. with the Buffalo Bills, Billy Sullivan with the Boston Patriots, now the New England Patriots, and Chet Sodom of the Oakland Raiders, which are now the Las Vegas Raiders.

Hunt and Adams weren't the only oil magnates in the Foolish Club. Barron Hilton, son of hospitality king Conrad Hilton (grandfather of Paris Hilton), made part of his fortune from his MacDonald Oil Company.

The New York Titans were only owned by Harry Wiseman for the first three seasons before financial trouble forced him to cut his losses and sell the franchise to a five-man syndicate of investors including Leon Hess, founder of the Hess Corporation (a large oil and gas company). Hess eventually became the sole owner of the team. By the mid-1960s, half of the AFL teams were run by oil tycoons.

Merger discussions between the AFL and the NFL began in the mid-1960s. When it was agreed that the winners of the two leagues would compete in an annual, final championship match, Hunt gave the event its name. In a letter to NFL commissioner Pete Rozelle dated July 25, 1966, Hunt wrote, "I have kiddingly called it the 'Super Bowl', which obviously can be improved upon."

Half of the modern NFL's teams were either founded or owned by oil tycoons in the 1960s. But when most people think of teams being owned by oil tycoons, there's one team that comes to mind. The Dallas Cowboys are the only team other than the Kansas City Chiefs to be founded and owned entirely by families who made a fortune in oil. The Cowboys were founded as an expansion team to the NFL in 1960, by oil man Clint Murchison, Jr. In 1985, the team was bought by H.R. Bright – also an oil man. Four years later, Bright sold the team to its current owner, Jerry Jones, for an unprecedented $140 million (£110 million). Though most of Jones's net worth is now tied up in the Cowboys, he originally made his fortune drilling wells in Arkansas, Oklahoma, and Texas.

As of April 2020, the Hunt family still own the Kansas City Chiefs and they remain one of the wealthiest families in the US. With superstar quarterback Patrick Mahomes on their roster, Super Bowls 54 and 56 brought them their first championship rings since that first Super Bowl back in 1969. The team, and the

Hunt family, who also own FC Dallas in Major League Soccer, are still as relevant as ever in the world of sport.

And the list goes on, through just about every sport. In the US the baseball team the Kansas City Royals is owned by John Sherman, who made his money in oil and gas. In ice hockey the Calgary Flames are owned by N. Murray Edwards, who made his money in Canadian Natural Resources. New York FC is owned by City Football Group (the Saudi ownership group that owns Manchester City FC in the Premier League, Girona CF in La Liga, and ES Troyes FC in Ligue 1). Sport is in bed with Big Oil. And it stretches beyond ownership.

Most of the equipment used in sports is made from plastic or other non-recyclable petro-based products, like carbon fiber (which is made from polyacrylonitrile, which in turn is manufactured from polymers and acrylonitrile – petrochemical products).

For instance, you wouldn't have modern sailing, rowing, canoeing, cycling, skiing, tennis, bobsleigh, luge, skeleton, golf, or ice hockey without carbon fiber in the main piece of equipment – the boats, bikes, racquets, sleds, clubs, and sticks. Carbon fiber is not recyclable and in many cases can't be repaired. This leads to unimaginable piles of waste emanating from the sports sector. Google "bike graveyard in China" for some grim images of mountains of waste (sorry in advance).

It's also hard to find a sport that doesn't include plastic somewhere in the supply chain. Most balls are made from plastic. Whistles for the referees. The running shoes. The uniforms. The signage around the sports venue, food service items at the concession stand, merchandise sold at events, you get the idea. But we forget, because it's not visible. And we ignore, because it's not pleasant. Most petrol products don't come with a handy sticker on them, announcing their origins. (Can you imagine? So many items would lose their appeal.)

Team ownership in sports is largely invisible, hidden away in a front office. We hear their names occasionally and they hold a ton of sway in the sports world. If they're lucky enough to see their team win a championship, the trophy will be handed to the owners first. But sport isn't about them. It's about the product on the field, the court, the ice, the pool, the track, the hill, or the road. Which is why sponsorship is so incendiary. Sponsor names are on the building, the jersey, and the sidelines. Sponsors, unlike owners, are center stage.

Why sponsor sports?

In 2020, the global sports sponsorship market was estimated to be worth $57 billion (£45 billion), with a compound annual growth rate of 6.72 per cent. In 2021, Nielsen Sports tracked a 107 per cent increase on sponsorship sales over 2020, marking a near-immediate bounce-back from the COVID lull. By 2027, Brand Essence Market Research projects the sponsorship market will reach $89.6 billion (£71 billion), roughly 12 per cent of the total value of the sports sector.

Sponsoring sports is an attractive marketing proposition for clothing brands, airlines, automotive companies, alcoholic beverages, sports drinks, and the list goes on. It's easy to see why: there are few other opportunities that allow corporate entities to so clearly align themselves with athletes, teams, and events that are so deeply beloved by fans, and the good values we associate with sports – teamwork, health, wellbeing, community, excellence. Academic and market research has consistently shown that sponsorship increases brand awareness, improves brand reputation, and can hedge against any bad press the company may attract for other activity.

It's shocking how effective this strategy is – so simple, yet so insidious. Unsurprisingly, much of the PR and marketing strategy associated with sponsoring the things we love – sports, the arts, education – were pioneered and perfected by – you guessed it – Big Oil.

Amy Westervelt is an American environmental journalist who has been investigating and reporting on the oil sector's climate research, denial, and lobbying efforts for more than two decades. Through this work, she's traveled to several oil industry conferences, launches, and events. She's visited the Exxon archives at the University of Texas, and has interviewed countless industry insiders and historians. Her research has unpacked the early PR work by oil companies to enmesh themselves in communities, spending millions on these efforts as early as the 1920s.

Over a call in February 2023, she shared some of her findings with me: "Really early on, [the oil sector and their PR teams] started shaping how people think about oil. They started tying it to patriotism and also this concept of the good and easy life, very much tying oil to leisure time, oil to wealth, and oil to progress." In one example, she told me about a campaign by Standard Oil of New Jersey, wherein the company was sponsoring bicycle safety training in schools across New Jersey and distributing flyers to kids to take home to their parents, as a way to associate oil with safety – and with good old American freedom.

We still see this today. A century later, oil companies regularly run campaigns that promote their products as the key to the "good life" through energy for our homes and fuel for private vehicles. From 2021 to 2023, a new string of advertisements aired on television in the US and Europe reminding viewers of everyday items that are made from petrochemicals, from pet products to hair dye to footballs. "They're right back to their original messaging," Westervelt explained. "It's a classic story that they keep going back to, because it has pretty much always worked to remind the public that they're reliant on oil and so they shouldn't be critical."

Westervelt continued, "The other thing they started doing back in the 1920s and 30s was to shape the public's understanding of the environment as something that's over there. It's nice to look at, but not as important as having cheap energy." This is a very convenient worldview for oil companies, whose social license depends on the world turning a blind eye to the consequences of drilling, fracking, refining, transporting, selling, and burning their products. So in the early twentieth century, oil companies latched on to this configuration of the human–nature relationship and promoted it with all the vigor of modern PR campaigns.

As oil has expanded around the world, they've taken this messaging with them: humans are the most important, nature is ours for the taking, and oil is the product that helps us achieve the "good life". A key strategy to buy local favor when an oil company moves into a new market is to sponsor the local sports teams or arts programming. In some cases, this has been going on for decades. If you take a drive from Houston to New Orleans along Highway 10, you'll see oil refineries out the window for hours on end. As I traveled that route in 2022, I listened to the radio to get a sense of what's happening in the area. I must have passed two dozen signs and heard another dozen radio advertisements for oil-sponsored youth sports tournaments, jazz festivals, and sports facilities.

Westervelt told me that she's noticed the same trend internationally: "The first thing they do, in almost any country that they operate in, is figure out what the most popular local sport is, and sponsor it. It's the number one way to win favor with a broad swath of the public. It makes them seem down with the locals, and people love it." As Westervelt and her colleagues were working on a story about Exxon in Guyana, they discovered that Exxon's first move in the country was to sponsor the cricket team, and then sponsor television coverage of that team's games. Westervelt recalls that when Exxon did that, "you had all these people being like, 'Thank God for Exxon. Without Exxon, we wouldn't be able to watch cricket on TV.'" To my knowledge, there's no other way to generate that much support and good feeling toward a brand in such a short period of time. Oil companies have mastered the art of buying good reputations through sports and arts sponsorships.

Oil turns to sports

So it's clear why the oil companies want to sponsor sports. Especially as the last few years have seen them ousted from other cultural institutions like museums and universities. In London,

England, the National Portrait Gallery and the Royal Shakespeare Company cut their ties with BP in recent years, adding to a growing list of oil-ditchers that includes the Tate Galleries, National Gallery, National Theatre, and British Film Institute – all formerly oil-sponsored properties. The British Museum and the Science Museum, which both still have oil sponsorships, have been under increasing pressure from activists and protestors to drop those affiliations. Sports are fast becoming the last battleground of oil sponsorship.

What's less clear, as time moves on, is why sports brands are so happy to accept money from oil companies who sell a product so clearly detrimental to the future of our planet and to their playing fields. Cycling is the sport I come to on this question, because it's so heavily owned and sponsored by oil, and yet cycling is universally heralded as a sustainable mode of transport and a solution to climate change. It seems like an apt place to start.

In case you're not a cycling fan, let me offer a quick overview of how deeply entrenched oil is in this sport. France's TotalEnergies began sponsoring Team TotalEnergies in 2016. In 2019, UK chemical producer INEOS took over title sponsorship of Team Sky, providing an annual budget of €50 million ($54.2 million or £43 million) – making it the richest team in the sport. Bahrain and the United Arab Emirates both own teams, while numerous state-owned Kazakh oil firms sponsor Astana. Then there's the new addition: Uno-X, a fuel station chain in Norway that entered the Tour de France in 2023. Three of the last four Tour de France races were won by riders whose teams were sponsored by companies or countries with links to fossil fuels. The first big race on the calendar each year – the Santos Tour Down Under in Australia – has oil in the title.

Mara Abbott retired from professional cycling after the Rio 2016 Olympics, where she placed fourth, capping a massively successful career that also included two Giro Donne wins. Over the course of her career, she was outspoken about the lack of funding in women's cycling. When I asked her about sponsorship deals in cycling, she was pretty blunt in her response:

> There isn't that much money in the sport, especially if you compare it to other big sports like the professional team sports, and so there isn't this plethora of opportunities to get big sponsorships. As a cycling team, you're in a position where you have to take what you can get. If you were to take a stand and say, "No, I am not going to be on the team that is sponsored by Shell or INEOS or any of the others", well, okay, but does that mean that you're not racing next year? Because that's what's at stake.

Cycling teams draw most of their budgets from sponsorship, making them a target for oil brands because it's so hard to turn the money away. Unlike football, rugby, and other big professional sports, cycling can't make money on ticketing since the events are held on public roads over thousands of kilometers. Prize money is also limited. Sponsorship is the big opportunity to ensure income – and survival – for a team.

In 2022, British Cycling announced a new eight-year sponsorship deal with Shell. The announcement was met with a swift backlash from activists, cyclists, and the broader outdoor sports community in the UK. An open letter to British Cycling authored by Badvertising, a climate campaigning agency, offered a scalding denunciation of the deal and asked British Cycling to withdraw:

> That you should allow the good name of British Cycling to become a billboard for an oil company which has for decades lobbied against environmental action and sowed doubt and confusion around climate science has come as a shock to many people … Whilst appreciating the challenges of funding any organisation, we ask you, for the sake of future generations, and to demonstrate that you reject greenwashing and sportwashing by major polluters, to withdraw from your sponsorship deal with Shell.

In less than 48 hours, the letter garnered signatures from more than 1,000 cyclists and organizations across the UK and made national headlines. By a week later, the number of signatures had doubled, and stories began leaking of cyclists canceling their memberships with British Cycling. The CEO resigned shortly thereafter, but still the partnership stuck. A few months later, my colleague Dr. Russell Seymour invited James Young, Commercial Director for British Cycling, to give a guest lecture in our Sustainability and Leadership course. The invitation was essentially an opportunity to offer our students the other side of the argument: why stick with this sponsorship when there was so much backlash?

I wasn't in class that day as Dr. Seymour was delivering his lecture, but I did grab the recording after the session. Young spent 47 minutes explaining the decision, starting with an overview of where funding comes from in that federation (memberships and event licenses, sponsorships, and government support through the National Lottery and UK Sport), and how the organization has been under pressure from the National Lottery to increase its own self-sufficiency by increasing the proportion of funding that comes from commercial avenues. This, he explained, made it hard for the other stakeholders to come back and say no to sponsorship dollars, even from a brand they may not be totally delighted to partner with.

The other argument he gave in support of the partnership was that Shell is positioning itself to be part of the much-needed energy transition in the UK and that there would be support for British Cycling to improve its performance and tackle its own net zero agenda with Shell's support. In fact, Shell was already helping British Cycling to work out its carbon footprint.

The argument basically boiled down to money. Shell had money, so Shell was in.

The organization has lost some members, but fewer than they were expecting. As part of the vetting process, the commercial team at British Cycling calculated the anticipated membership losses that this announcement would bring, but Young was pleased to announce it wasn't as much of a loss as they had anticipated. He admitted that his team is aware that "around 70 per cent of the national population will probably not see the partnership in a good light and we probably won't change their mind about the partnership."

Andrew Simms, who wrote the open letter, is an author and leads campaign works with the New Weather Institute and Badvertising. He watched the sponsorship deal play out and believes it's part of oil's final surge of sponsorships in sport. Simms told me,

> What's been really interesting is the speed with which the issue has risen up in sport in just the last two years, which has happened in tandem with the climate movement generally gaining steam. There is undoubtedly rising momentum that cuts across sport, and international regions. Within sport, you've got all the same institutional and historical dynamics that you've got in any other sector. But there's so much more pressure on sport organisations now than there was a few years ago.

Simms is referring to the public outcry over the Shell sponsorship of British Cycling, the rage bubbling in the cricket world over Saudi Aramco's sponsorship of the International Cricket Council, and the increasing frequency with which protestors are showing up to shout down oil sponsors at sports events – from the Santos Tour Down Under to the Tour de France. The notion that oil should not sponsor sports is not yet totally mainstream, but it's getting there.

In case you've read this far and aren't totally sure why I'm critiquing oil sponsorships: oil companies are bad actors. Recent investigative journalism – of which Amy Westervelt was a big part (and I strongly encourage you to listen to her podcast, *Drilled*) – has revealed that companies like ExxonMobil, Shell, and BP have known about climate change since the 1970s, realized this would be bad for business, and proceeded to bury the science, muddy the waters on climate action, mislead the public on the safety of their products, and promote climate delay for five decades. This isn't conjecture; internal documents from these companies, and speeches at company meetings and industry events, say as much.* A growing body of academic literature confirms it.

More recently, oil companies have shimmied their way out of countless major environmental and social justice issues. One example is the *Deepwater Horizon* oil spill in the Gulf of Mexico in 2010, which was still not fully resolved a full decade after its occurrence. Another example is the court case ongoing between Chevron (who took over for Texaco when they bought the company in 2021) and 30,000 Ecuadorian Indigenous people who sued the oil giant to seek compensation for the 30 billion gallons of toxic waste and crude oil that was dumped in the Amazon rainforest, poisoning their water supply between 1972 and 1993. It's clear that Big Oil doesn't want to be charged for the pollution it has caused globally, nor for its contributions to climate change.

The reason oil companies have been able to get away with decades of false advertising and steamroller their way into winning lawsuit after lawsuit is their wealth. Oil companies are ridiculously wealthy. In 2018, Global Justice Now compiled a list of the 200 wealthiest economies, including countries and corporations: 69 of the 100 richest economies are corporations; ten of them are oil companies. These same companies show up, unsurprisingly, on the Carbon Majors Report – a list of the 100 companies responsible for 71 per cent of global emissions since the 1970s. These companies are: Gazprom, National Iranian Oil, BP, ExxonMobil, Royal Dutch Shell, China National Petroleum

---

*The Union of Concerned Scientists and Greenpeace have done a wonderful job of collating evidence of corporate disinformation efforts by fossil fuel companies. One of the more complete collections is "The Climate Deception Dossiers", published in 2015 and easily found online at https://www.ucsusa.org/resources/climate-deception-dossiers

Corp, Chevron, Total, Lukoil, and Saudi Aramco. My point is this: oil firms globally are making more money than most of our countries by polluting us all. Because of this, they don't deserve to be celebrated and associated with our favorite sports brands.

Now, it's true that we can't just divorce Big Oil completely and turn the oil taps off tomorrow, because renewable energies – wind, solar, hydro, fusion, fission, and so on – can only cover a small percentage of our needs *at the moment*. However, we can reduce our use of fossil fuel energy quickly by doing all those small things that reduce the energy load in our homes: turning off lights, installing better insulation, switching to heat pumps, and so on. We can also drive less and fly less (especially short-haul flights that can be easily converted to train or bus travel, and business travel that can be eliminated and substituted by Zoom). We can encourage governments to stop subsidizing fossil fuels, which are one of the most heavily funded industries, and instead invest those public dollars into renewables. And in the context of sport, we can stop them from continuing to promote a toxic product that is hurting people and the planet on the jerseys, buildings, event posters, and advertisements linked to sport.

Being front and center on our sports programming is a privilege and I don't think it's ridiculous to suggest we should disallow unhealthy products to promote themselves through sport – in fact, there's a good precedent for kicking out fossil fuels. An army of doctors, lawyers, and activists waged this war once before, and kicked out tobacco.

Tobacco wars

Let's step back in time. In the 1900s, tobacco was widely viewed as a healthy product. Doctors would encourage people to smoke to clear their lungs, believing that by coughing, they were bringing up all the bad stuff in the airways – as opposed to responding to the poison in the cigarette. Through the mid-twentieth century, smoking became a status symbol, the icon for "cool", and started to show up in every aspect of pop culture: television, movies, and of course sports. When several countries passed legislation banning cigarette advertising on television in the late 1960s, tobacco companies shifted their money from producing commercials to sponsoring live sports coverage. This way, they could be seen on television on the sidelines or on the jerseys of athletes for the full two hours of the event, rather than the 30-second commercials during breaks, while still complying with the law. A brilliant (terrible) loophole.

In the 1970s, you could turn on your television to watch the Virginia Slims Women's Tennis Circuit, the Marlboro Cup horse race, and the Winston Cup NASCAR (stock car racing) series. By 1990 more than 20 different televised cigarette-sponsored sports existed in the USA alone, including eight different forms of motor racing. With just two exceptions out of 28 teams (at the time), every Major League Baseball stadium featured prominent outfield billboards for either Marlboro or Winston cigarettes. A similar scenario existed in the National Football League and National Basketball Association.

Throughout the second half of the twentieth century, however, it became clear that it was illogical to use athletes and sports events to promote a product that, according to the World Health Organization, kills more than 8 million people each year. A series of anti-tobacco campaigns popped up in several countries, led by doctors, moms, lawyers, and sports fans. For the most part, they were successful, throwing enough shade on tobacco to inspire some sports events to drop their tobacco sponsors voluntarily. These campaigners also lobbied for legislation that would ban the product from advertising altogether. The 1984 Summer Olympics in Los Angeles were the last Games to feature an official cigarette sponsor and marked a turning point globally in tobacco sponsorship.

The first country to implement a ban on tobacco advertising was Norway in 1975. This ban prohibited all forms of tobacco advertising on television, radio, billboards, and in newspapers and magazines. The European Union (EU) followed suit in 1991, with a directive that banned tobacco advertising on television and radio, and a subsequent directive in 2003 that extended the ban to print media and sponsorship of events. In 1998, the World Health Organization (WHO) Framework Convention on Tobacco Control (FCTC) was adopted, which called on countries to implement comprehensive bans on tobacco advertising, promotion, and sponsorship. As of 2021, 182 countries have ratified the FCTC.

However, not all countries have imposed the same types of restrictions. Many countries have implemented comprehensive bans, including Australia, Canada, France, Ireland, New Zealand, Norway, South Africa, and the United Kingdom. Others, like the United States and Japan, have opted for partial bans, such as banning tobacco advertising on television and radio, and in sports sponsorships, but allowing it in print media. Still, this made an important dent: one study out of the US estimates that the total sponsorship spend by tobacco companies in 1992 was $136 million (£107 million) across all sports. By 2013, three years after the full ban took effect, that number was down to zero. In places where the industry is more powerful and more people smoke, like India and China, the bans are quite limited, extending only to television.

These holdout countries, the ones that have not implemented complete bans, are part of the reason that some sports properties have not totally kicked the tobacco habit yet. In 2020, a report by Stopping Tobacco Organizations and Products (STOP) found that in the product's 70 years of sponsorships of Formula One, the sport took in more than $4.5 billion (£3.5 billion) from tobacco companies, with more than $100 million (£80 million) coming during the current 2020 season. So the bans haven't managed to get tobacco completely out of sport, but they have managed to drop sponsorship considerably across most sports. More importantly, the legislation has reduced the visibility of tobacco brands in sport, with most sponsorship now happening through subsidiary brands, not a pack of cigarettes on a billboard on the side of the racetrack.

Taking the legal route

Activists are hoping that similar legislation will eventually limit the advertising options available to oil and gas, including through sponsorships. Already in France, a new piece of legislation was adopted in 2022 that will prohibit advertising for all energy products related to fossil fuels, including petrol products, coal energy, and hydrogen-containing carbons. A year earlier, Amsterdam became the first city in the world to prohibit ads from fossil fuel companies and the aviation sector. The Amsterdam ban is stricter than the French policy and includes petrol-fuelled cars and flights, removing all images and advertising for fossil fuels city-wide. Sydney's deputy mayor Jess Scully has announced plans to follow suit with a similar advertising ban there. But it feels as if this movement is not going fast enough.

A 2022 investigation into the Australian sports sponsorship market found 51 sponsorship deals between fossil fuel companies (oil, gas, coal) and sports properties, across Australian Rules football, rugby union, rugby league, cricket, soccer, and netball. Professor Emma Sherry, lead author of the study, said in the press release, "These sponsorships amount to around $14 to $18 million – a number that, while not small, could be replaced with less harmful options in the coming years."

In some cases, sports organizations are dropping their fossil fuel sponsors, citing misalignment with their strategic sustainability goals. A standout example is the American Birkebeiner – more commonly known as the Birkie – North America's biggest cross-country ski race, which announced it was ending its sponsorship deal with Enbridge, a Canadian pipeline and energy company. In a letter to their community, posted on the Birkie website, the organization said: "We've taken pause to reconsider our relationship with Enbridge Energy and have chosen to dissolve our agreement." It continued, "In hindsight, we realize that this association was perhaps not a clear pathway to engaging conversation in support of education, future change, and ultimately our greater Birkie Green initiatives."

The move was supported by skiers but not by everyone. I spoke to Ben Popp, Executive Director of the Birkie, about that decision. "The skiers got it," Popp told me, "but most of our skiers come from outside this area, they come for the race, for the competition, and then they leave. And they're noticing the changes in the winters so they know that climate change is a concern. In that group, the decision was well received. But locally, we're in a pretty conservative area in northern Wisconsin. So, from a social media standpoint, and then just the general media, I was in the newspapers for a while and not in a good way. That wasn't fun."

In 2021, Tennis Australia announced they were dropping Santos as their Natural Gas Partner, eliminating the category altogether after significant backlash and accusations of greenwashing. A petition was circulated and gained more than 7,500 signatures against the partnership. It became too much for Tennis Australia to ignore, so they dropped Santos. Great work, 10 out of 10.

A few months later, when the Russo-Ukrainian war began in February 2022, UEFA dropped Russian state-owned oil company Gazprom as its sponsor, releasing its contract worth €40 million ($43 million, £34 million) per year to distance themselves from Russia. Sure, this was done for political reasons linked to the war, not concerns over the oil company itself, but it had important green benefits: it made UEFA a fossil fuel-free sports property.

But still, all these actions are voluntary. These are the exceptions. It hasn't yet become a trend for sports organizations to drop oil sponsors.

Some activists put high-emitting industries like airline and automotive companies in the same boat as oil companies. This includes the Rapid Transition Alliance, which released the Sweat Not Oil report in 2021, tracking the number of sponsorship deals in sports across all high-emitting industries. They found 258 sponsorship deals, and that didn't even include all sports.

Lew Blaustein, founder of Green Sports Blog and the non-profit EcoAthletes, has been vocal about the need to drop all high-emitting sponsors:

> I believe that, if sports leagues and teams really want to demonstrate leadership on climate, they need to hold their sponsors to a much higher standard than they do now. What would that look like? Ban the extractive industries for sure, they're the tobaccos of climate. But that's the easy part! If other high-emitting industries, like big chemical, big automotive, and big ag want to continue to sponsor and advertise on sports, they should show that they are on a path to real emissions reductions.

Amy Westervelt agrees.

> If you go back in history, it's clear that a few big PR companies came up with strategies to buy public favor and extend the social licenses of all the industries that were either fouling the commons or using the commons for their own benefit, but not actually paying the public to do so. And they deployed those strategies for all of them, not just oil. So they all do it. And I don't really see oil as being any worse than any of the other industries. I actually think it's weird that the automotive industry has not gotten called out a bit more on this. Because they are terrible.

A few years ago I sat on stage at the Green Sports Alliance Summit in Philadelphia and made this same case. My argument – which I still stand by – is that it's confusing for fans when you show up to a match with a "Green Football Weekend" slogan attached, or an NFL Green campaign going on, and you look around the stadium and see logos for airlines, auto, and oil companies in every line of sight. It sends mixed messages about the sport organization's commitment to sustainability and the wellbeing of future generations. Consistency with climate messaging is key. Big Oil advertisers blur that message.

I'll let you make up your own mind about whether all high-emitting industries need to go or just oil. But I'll put a flag in the ground and say the oil industry in particular is toxic and should be immediately kicked out.

A shift away from fossil fuels in sport is aligned with the global conversation on climate change. We're currently going through a 20-year process of economic change. And however fast or slow the curve bends, it is bending only in one direction, and that's away from a fossil fuel economy and toward a green economy. As that happens, the businesses within the green energy sector will be shifting their investments too, so we may see a whole range of new entrants into the sponsorship game from solar, wind, regenerative agriculture, and all sorts of industries that are currently nowhere to be found in sport. By 2040, we could see a wide variety of sponsorships in sport that represent a different set of values. Maybe sooner. Ideally now.

CHAPTER SIXTEEN

THE ROSTER: ATHLETES VS CLIMATE CHANGE

In early 2023, a 17-year-old cross-country runner made international headlines when she sent a letter to British Athletics asking not to be considered for the World Championship team as she would not travel to Australia. Innes FitzGerald, a leading junior endurance runner, cited her "deep concern" over the carbon emissions of travel as her reason for bowing out.

In her letter, published by *Athletics Weekly*, she said: "When I started running, the prospect of me competing in the world cross country championships would have seemed merely a dream. However, the reality of the travel fills me with deep concern." She goes on to say, "I would never be comfortable flying in the knowledge that people could be losing their livelihoods, homes and loved ones as a result. The least I can do is voice my solidarity with those suffering on the front line of climate breakdown."

This wasn't the first time FitzGerald had taken this position. Just a month earlier, she had placed fourth in the European Cross Country Championships against much older and more experienced runners. After the race, it was discovered that she had spent days in transit getting to Italy – a combination of buses, trains, and a folding bicycle to shuttle between stations.

FitzGerald is part of a generation of athletes that will force us to rethink the way sport operates. But she's not the first. She joins a long and growing list of athletes on the frontlines of climate change, experiencing its impacts in their sports and witnessing climate hazards hurting communities around the world as they travel to compete. Some have been quiet in their protest, others loud, with the chorus reaching a decibel level that will soon be hard to ignore.

Athletes take the lead

The story of athlete activism stretches back over a century. Jesse Owens, Wilma Rudolf, Jackie Robinson, Tommy Smith, John Carlos, Althea Gibson, the 1994 South African national rugby team, Kareem Abdul-Jabbar, Venus and Serena Williams, Colin Kaepernick, and Akim Aliu, are just some of the athletes who have put their careers on the line, risking their finances, their playing opportunities, their sponsorships, and their reputations for Black liberation. A similarly long list can be drawn up for women who have fought for gender equity in sport and beyond: Gertrude Ederle, Kathryn Switzer, Billie Jean King, Allyson Felix, the US women's national football (soccer) team, and more recently, the Saudi women's national football team. Right now in Kenya, a team composed of champion marathon runners, including former world record-holder Mary Keitany, has assembled under the banner "Tirop's Angels" to protest about violence against women in their community and to educate young people and women about healthy relationships.

Athlete activism works because athletes are highly visible and influential. In some cases, individual athletes have larger platforms than the teams or leagues they belong to. The Pantheon Project by the MIT Media Lab examined which figures were the most memorable and socially influential in society through time. From AD 1 to 1450, it was politicians and religious figures. At the close of the Middle Ages, through the Renaissance, the Enlightenment and into the early Industrial Revolution, roughly 1450 to 1880, writers, artists, and scientists moved into top roles alongside politicians, who maintained their clout. By the late 1880s, actors, musicians and singers joined the top tier of influential figures. As professional leagues and media coverage of sport ramped up in the twentieth century, athletes were propelled to the top spot as the most memorable people and have stayed there.

Most research into athlete activism has demonstrated that this visibility is a key ingredient for athlete activism. And yet not all athletes use the platform. It can be scary to speak out and the pushback can be ruthless. Billie Jean King and the women of the original Women's Tennis Association signed one-dollar contracts, effectively giving up their place in the men's tennis world to establish something new, with no guarantee of success. Despite being the highest scoring player in the NBA at the time, Kareem Abdul-Jabbar was vilified in the media for his social justice activism addressing racism, Islamophobia and other forms of discrimination. In 2016, Colin Kaepernick lost his job as a star quarterback over his peaceful protest during the US national anthem. As recently as 2018, LeBron James and Kevin Durant were told to "shut up and dribble" by Laura Ingraham of Fox News, over their criticism of President Trump. James responded on social media, posting #wewillnotshutupanddribble. The line became emblematic of the public vitriol athletes face over their activism, in a society that wants to see athletes perform their roles as entertainers and nothing more.

The year 2020 was a turning point for athlete activism. The *Toronto Star* called it "a watershed moment". The *New York Times* dubbed it the year when "even sports wasn't an escape". And *Sports Illustrated* made athlete activists the "Sportspeople of the Year", naming LeBron James, Naomi Osaka, Breana Stewart, and two players on the Super Bowl-winning Kansas City Chiefs roster, Patrick Mahomes and Laurent Duvernay-Tardif, as icons of a growing movement to bring about change through sports. In June 2020, in the height of the pandemic and mere weeks after the murder of George Floyd at the hands of Minneapolis police, Megan Rapinoe, Sue Bird, and Russell Wilson opened the ESPY Awards with a powerful video about athlete activists through the ages. The video concluded with a call to action: "our return must be part of the fight for justice ... our return is our turn to stand up for what's right."

The call was answered. Weeks after the ESPYs aired, the Milwaukee Bucks of the NBA and the Milwaukee Brewers of the MLB announced they would not be playing their games after the police shooting of Jacob Blake in Kenosha, Wisconsin, just 40 miles from their home venue. Over the course of a few hours, other teams joined them in striking from games: six teams in the NBA, seven teams from the NFL, eight teams in the MLB, nine teams in Major League Soccer, and Naomi Osaka. The National Hockey League canceled all games for a day during playoffs in solidarity with the athletes in other sports. It was a remarkable show. All sports just... stopped. Again. After the COVID shutdown and re-opening, athletes chose to shut it down once more, over morals.

In those days of widespread strikes, North American sports fans were forced to sit uncomfortably with the notion that sports are a privilege, not a given. If society isn't healthy – physically or socially – sports will stop. It was also a stark reminder that athletes are people first, and sometimes it's more important to grieve and be human, rather than to play a game for others' entertainment. Many Black athletes expressed the view that the police murders of Black people are an unwelcome reminder of their own vulnerabilities in a systemically racist police system.

———

The year 2020 reminded sports fans that athletes are people, too, and that sports are not apolitical. They never were. After all, we sing the national anthem at professional matches, send our top athletes to visit the White House or get a medal of honor from the king, and cheer on national teams in international competitions – one of the most visible expressions of nationhood in modern society. As people woke up to the reality of athletes being real people with real concerns about the world, support for athlete activism grew.

A *Washington Post* poll conducted in 2016 and repeated in 2020 found that in that four-year span, public support for athlete activism jumped from 20 per cent to 62 per cent. A 2023 survey of the British public by UK Sport found two in three UK adults (66 per cent) believe that athletes have a role to play in championing causes they believe in and raising awareness of social issues.

A new angle

With the backlash from the public still very present but waning, a new form of athlete activism has entered the fold in recent years: climate activism. This one is a little different from other forms of activism in that it's not always clearly linked to an individual's identity markers. Unlike in the equal pay battle of the feminist movement, or the police brutality protests of the Black Lives Matter movement, it is not always straightforward for an athlete to articulate the reasons they affiliate themselves with the cause of climate change. It is easy for fans to see and understand why a female athlete might protest for equal pay, or why a Black athlete would relate to and support the Movement for Black Lives. This is not to discount the importance of allies and mutual support – those are critical. My point is that athletes who are advocating for issues that match their visible identities do not have to explain their "why". Climate activists often do.

At this point in the book, it should be clear why an athlete would be worried about climate change. This ongoing manmade disaster is wreaking havoc on lives and livelihoods all over the world. Athletes have a front-row seat to the destruction, experiencing heatwaves, shorter winters, heavy rains, drought-stricken fields, and polluted air and water first-hand. However, it's not easy to communicate these issues in a tweet or an Instagram post.

The biggest barrier – the one I hear about most often when I speak to athletes about their activism – is the hypocrisy trap.

This typically shows up as a quip by social media trolls and media personalities who point to athlete activists and say "Well, you travel the world and live the high life, you're part of the problem, so you shouldn't be talking about climate change." It's such an easy line, and athletes fall victim to it. This is exactly the same kind of criticism that politicians and scientists experience when they talk about climate change. Even Greta Thunberg gets called a hypocrite.

Here's the thing: nobody is perfect when it comes to climate action. Until FitzGerald came along in 2023, no athlete had actually passed up championship opportunities because of a moral quandary with flying, and it's unlikely to happen much more – because the sports sector depends on athlete travel. We either have some amount of travel (which can definitely be reduced) or we don't have competitive sport. Athletes travel because it's their job to travel, in the same way it might be your job to be at the office at 9am, and if you live in a place without good public transit, you're probably driving to work. If we want to continue to watch sport, we need to accept that athletes travel. Sure, they can make some more environmentally friendly choices elsewhere in their lives – switching to a vegan diet, driving an electric car or a bike, making sure their investments aren't funding fossil fuels – but those are items they may actually already be doing, it's just less visible. It boils down to this: nobody is going to be perfect, not even athletes. It's an impossible expectation, and an unfair one.

My friend Jeremy Casebeer, a ten-year veteran on the AVP beach volleyball tour, has a great response to the hypocrisy trap. He said,

> If you look at the footprints of business leaders and business travel, athlete travel is tiny in comparison. But we're way more visible, so we get called out ... I'll get people in the comments section accusing me of being a hypocrite, but I can't change all of that, it's my job, it's how I pay the bills, the league calls the shots on travel. What the trolls on social media don't

is what I'm doing behind the scenes. I'm personally offsetting my travel, I'm working with different environmental projects that I care about, I'm serving as an ambassador for tree planting efforts and Parley for the Ocean … and when I lay that out for people and be honest about what I'm doing, they don't have much of a response.

George Monbiot of the *Guardian* sums it up nicely: "Hypocrisy is the gap between your aspirations and your actions. Greens have high aspirations – they want to live more ethically – and they will always fall short. But the alternative to hypocrisy isn't moral purity (no one manages that), but cynicism. Give me hypocrisy any day."

The other challenge climate activists face is do-gooder derogation. When a person adopts an environmentally friendly behavior, like switching to a vegan diet or buying an electric car, others may see this and start to feel defensive about their own shortcomings. This defensiveness leads the others to pick apart the do-gooder's behavior and intentions, and reject the exemplar altogether. Do-gooder derogation can happen subtly or more overtly, ranging from a quip by a teammate in the locker room to a dismissal by broadcasters on national television. When Lewis Hamilton speaks about his vegan diet on social media, he's often met with a mixed response of some support and some criticism.

Slowly, things are changing and the challenges of backlash, hypocrisy trap, and do-gooder derogation are becoming less important. Part of this can be chalked up to a growing awareness that climate change is real and urgent, and a growing segment of the population is concerned. Athletes, too, are concerned. A 2021 survey by World Athletics found that 77.4 per cent of elite track and field athletes are either "very concerned" or "extremely concerned" about climate change. A similar survey run by World Biathlon found that 90 per cent of biathletes feel their sport has already been impacted by climate change. With growing concern come growing numbers of activists.

Climate activism by athletes has taken many forms. One of the earliest climate activists in sport was NHL all-star Andrew Ference, who partnered with Canadian environmentalist David Suzuki in 2007 to develop a carbon offsetting program for NHL players. After most of his teammates on the Calgary Flames signed on in year one, Ference took the initiative league-wide and garnered participation from over 500 players by year three. In 2012, National Geographic aired a ten-episode Web series called *Beyond the Puck* about Ference's work in the environmental movement. Since retiring in 2018, he's been working at the league as their inaugural director of social impact, growth, and fan development, and he continues to be involved in environmental work through the league's NHL Green program.

For some retired athletes, protesting is the action of choice. In 2019, former Paralympian and double-gold-medalist James Brown scaled an airplane at London City Airport to demonstrate against the damage caused by flights. He spent an hour atop the aircraft, live-streaming his protest on Facebook, before firefighters forcibly removed him. His action managed to shut the airport for a whole morning, disrupting more than 300 passengers and costing the airline £40,000 ($50,000). At his trial two years later, Brown wept as he told jurors: "I was prepared to challenge myself, to be scared, to face the fear, because the fear of climate ecological breakdown is so much greater." He was found guilty of causing a public nuisance and was sentenced to 12 months in prison, of which he served half.

Brown's high profile as an athlete was part of what made the action successful, garnering international headlines. So when Etienne Stott, Olympic champion in canoeing at the London 2012 Games, decided to participate in a similar action in April 2022, he knew his visibility as a public figure would help the demonstration succeed. Stott and four other protestors climbed onto a Shell oil tanker as it tried to leave a petrol station in west London and hung a banner reading "End Fossil Filth". The protestors then glued themselves to various components of the tanker to prevent it from moving. It took a specialist team of police de-bonding experts to remove them. In court, the judge dismissed the charges of tampering with a motor vehicle, stating that the oil tank itself was not a motor vehicle. A technicality – but an important one, which made it easier for the judge to let them go.

Other athletes are working behind closed doors and within their organizations to effect change. Gaby Dabrowski, a top ten-ranked doubles tennis player, has been agitating for change within the WTA for several years, especially around plastic waste on the women's tour. Jeremy Casebeer, whom I mentioned above, chose to work with the AVP rather than against them as he co-developed the first sustainability strategy in the sport of volleyball. In golf, Aubrey McCormick leveraged her position as an athlete in the LPGA to support a number of courses to adopt more sustainable management practices – this was more than a decade ago, before sustainability was "cool".

Awareness-raising is perhaps the biggest-ticket item for athlete activists. Perhaps you've heard of Lewis Pugh, who has been swimming in some of the most remote parts of the world to raise awareness of ocean pollution and ocean acidification. Or maybe you've come across Will Gadd, an ice climber who has been sharing stories of lost glaciers in different parts of the world. There are also several voices coming out of the snow sports movement, particularly in skiing and snowboarding. Protect our Winters, a non-profit launched by snowboarder Jeremy Jones in 2007, now has hundreds of members in its athlete alliance who compete in these sports and use their platforms to advocate for important political action on climate change.

Climate activism shows up in a number of ways. David Pocock was born in Tanzania and moved to Australia as a teenager, where he played rugby competitively and eventually made the national team in 2009. For 11 years, Pocock competed internationally, traveling the world in the yellow and green jersey, wearing an airline logo on his chest. During his playing career, he vocally supported government policies on carbon pricing and participated in non-violent protests at the Leard blockade against the expansion of the Maules Creek mine in the Leard State Forest – for which he was arrested. When he retired in 2020, Pocock dedicated his time and career to conservation and climate action. He was instrumental in leading the Cool Down initiative in 2021, which saw 450 Australian professional athletes sign a letter to the Australian public urging them to act on climate. A year later, he ran for a senate seat as an independent and won, making him the first independent senator from the Australian Capital Territory. In 2022, he was named Athlete of the Year in the BBC's Green Sports Awards.

Sebastian Vettel, a Formula One driver, famously left the sport in 2022 because, as he put it, "we have another race to win". Also in motorsport, Lewis Hamilton is an outspoken advocate for climate. And while he gets attacked for flying around the world and being a driver (remember the hypocrisy trap), he also owns a team in Extreme E, which is an off-road electric vehicle racing competition that holds races in some of the most heavily impacted parts of the world to draw attention to these regions and the challenges they're facing. On social media, Hamilton has been vocal about his concerns over climate change, and has committed to a plant-based diet.

Also on the vegan train, you have current and former athletes across a range of sports: tennis stars Novak Djokovic and Venus Williams, soccer players Hector Bellerin, Alex Morgan, and Jermain Defoe, NBA all-stars Chris Paul and Kyrie Irving NFL players Cam Newton, Tony Gonzalez, Colin Kaepernick, and Tom Brady, and of course all of the mixed martial artists behind the documentary *The Game Changers*, which aired on Netflix in 2018 and earned a score of 99 per cent approval on Rotten Tomatoes.

Garry Gilliam, another vegan athlete, is a retired NFL player who has been using his platform to create opportunities for people in his hometown, the under-resourced community of Harrisburg, Pennsylvania. Through his non-profit, The Bridge,

Gilliam has created a space for learning and co-working, a hydroponic farm to produce healthy foods for his community, and is in the process of setting up a program that will improve people's financial literacy and employability. His work sits right at the intersection of environmental and social issues, which he views as the key to success.

Gilliam's initiatives are designed to target and undo systematic forms of oppression in his hometown – both social and environmental. He admitted to me that it's hard work and likened it to fighting a supervillain.

> We say we're assembling the Avengers, if you watched any of that stuff. So, who's our enemy? Thanos is our enemy. Who's Thanos in the real world? That would be systemic oppression and systemic racism. The systems that have been designed and implemented strategically in a lot of different places, not just the United States, to prevent and slow economic inclusion, food security, and social welfare. In the news, we hear about food deserts, but at The Bridge, we call it food apartheid. Deserts naturally occur, and we know that healthy foods being made unavailable in certain areas is very strategic. That's really what The Bridge is about. It's about putting together a system that combats those injustices.

Gilliam's work has propelled him to becoming one of the most visible and active environmental advocates in the sport of American football.

In the last few years, we've seen a sharp growth in the amount of coordinated, organized climate action by athletes. In 2021, David Pocock (Australia), Alena Olsen (USA), and Jamie Farndale (Great Britain) – all elite rugby players who compete on their respective national teams – solicited signatures from more than 200 professional rugby players in an open letter to World Rugby, urging the federation to act on climate. Weeks later, the three organizers were invited into meetings with World Rugby officials to consult on the development of rugby's first sustainability strategy, launched in early 2022.

In 2022, Australian cricketer Pat Cummins launched Cricket for Climate, a collective of cricket athletes – with some support from soccer players and others – who are raising funds and organizing solar panel installations on cricket clubs across Australia to promote green energy and reduce the emissions linked to their sport.

In October 2022, Cummins, who captains the Australian men's national cricket team, publicly voiced concerns over energy company Alinta's sponsorship of Cricket Australia. He told the media, "When we're getting money, whether it's programs for junior cricket, grassroots, things for fans around Australia, I feel a real responsibility that with that, we're doing on balance what is the right thing." The Alinta sponsorship, worth Australian $40 million ($50 million or £20 million) was later canceled, with both Alinta and Cricket Australia emphasizing it was not because of Cummins' pressure. From the outside, it's hard to see how the star athlete's strong dissent didn't factor into the decision.

In an email, Cummins told me,

> I don't consider myself a climate activist at all. I'm a father, a husband and a cricket player who wants to make a contribution and to leave this planet in better shape than it is today. Like everyone, we are all finding our way through this transition to a more sustainable future and if I can take some practical steps in my own life and also support our Cricket community to do the same, then I'd be pretty happy.

The same week, Donnell Wallam on the Australian Diamonds netball team objected to wearing the uniform with Hancock Prospecting logo on it. The Hancock deal, worth $15 million ($9.5 million or £7.5 million), had been announced just one month earlier and represented an opportunity for netball to secure its financial future after years of tight finances. Wallam's concerns were less about climate and more about the Hancock ownership's historical racism. Billionaire Gina Rinehart, owner of Hancock Prospecting, is Australia's richest person. Her family made their money from mining, with little concern for the environment and even less concern for the Aboriginal people of Australia. In 1984, Rinehart's mining magnate father, Lang Hancock, notoriously called for the sterilization of Aboriginal people. Wallam's teammates got behind her, refusing to wear the logo. The group effort was enough to spur nationwide media coverage and shut the sponsorship down.

Increasingly, we're seeing non-profit organizations popping up on every continent to facilitate athlete education on climate issues and convening groups of athletes to take action together. In the US, Players for the Planet was launched in 2008 by professional baseball players Chris Dickerson and Jack Cassell when they saw the impact of plastic pollution in baseball clubhouses. To date, they've engaged 200 professional athletes in a range of activites: beach clean-ups, e-waste recycling events, plastic reduction initiatives in baseball venues, sustainable upgrades to youth baseball fields, and conservation and restoration efforts in the US and beyond. Most notably, in the last few years, they've run an annual beach clean-up in the baseball-crazed Dominican Republic, led by MLB players, which collected 215lb of PET plastic, 80lb (97kg) of HDPE plastic, 45lb (20kg) of foam, 18lb (8kg) of Tetra Pack, and 20lb (9kg) of shoes in 2023. Reflecting on the progress to date in a video on social media, Dickerson said, "Having 150 people on the beach is great. But it's going to take people raising hell to really change the issue, it's going to take a lot of buy-in from a lot of individuals – high-ranking individuals – who are going to have to make a decision on what is best for the future."

In Britain, Melissa Wilson, former Team GB rower, and Olympic champion sailor Hannah Mills formed Athletes of the World to galvanize athletes in the Olympic movement to engage on climate change. Their first major campaign saw over 50 Tokyo Olympians and Paralympians from around the world, including flag bearers from 35 different countries, challenging World Leaders about the urgent need for ambitious climate action in a video that was launched on social media during COP26 in Glasgow in 2021. Since starting their work in 2021, Athletes of the World have run athlete education sessions for Formula E, Premier League teams, the athletes competing in the 2022 Commonwealth Games, and more. Theirs is a project based on outreach and empowerment – and it's working. In just two years, they've engaged more than 600 athletes in their campaigns and trained groups of elite and professional athletes from more than 20 sports, including some of the most visible athletes in the world.

The most exciting thread of the athlete activism work, in my opinion, is the potential for the next generation of elite athletes to mainstream climate action in sport. Inspired by Greta Thunberg and the Fridays for Future movement, young people are coming of age with powerful role models demonstrating collective action. In a survey study of 1,300 US adults, a team of British and American climate communications researchers found that Americans who report being familiar with Greta Thunberg also feel confident that they can help mitigate climate change as part of a collective effort. These people are also more willing to take action themselves.

Subsequent research out of Norway found that young people, particularly those who are already concerned about climate change, are inspired by Greta Thunberg to think differently about the systems that have caused the climate crisis. Taken together, these findings suggest we could see more climate activists coming up in the next generation. Researchers have dubbed this the Greta Thunberg effect.

EcoAthletes, a non-profit based in New York City, is betting on young athletes to be the next big segment of activists in this space. In the last two years, the organization, which supports elite and professional athletes to speak up on climate change through training and networking opportunities with scientists and climate activists, has shifted its gaze to the up-and-comers: college athletes.

Cam Bentley, a rower at the University of Virginia, believes the next generation of athletes is only getting started. "We are the first group to feel the impacts of rising temperatures and extreme weather events on our training and competition. Such disruptions have forced greater awareness, and subsequently, a call to action." The coming generation also benefits from role models. Bentley is particularly inspired by Lauren (Lu) Barnes, captain of Seattle's OL Reign in the National Women's Soccer League. Bentley tells me, "I think Barnes demonstrates the power of athletes to affect the operations of their organization, the behavior of their teammates, and the resources available to athletes." As the saying goes, if you can see it, you can be it.

This brings us back to Innes Fitzgerald, the 17-year-old who has gone one step further than athlete climate activists who have come before. Fitzgerald is not only voicing concern about climate change, generally. She's turning her gaze inward, focusing on sport itself as part of the problem. Actions like hers – and the media storm they stir up – might be exactly what the sports sector needs to force it to think more critically about its own footprint and make the big changes that are overdue. With Fitzgerald, the call is coming from inside the house.

CHAPTER SEVENTEEN

WHAT'S NEXT?

In 2021, a video circulated on social media that caught my attention. It begins with the unmistakable sound of crackling fire. As the image comes into focus, a voiceover in Finnish begins with subtitles at the bottom. "My grandfather always told me that Päivätär would come back one day. The Goddess of the Sun." The narrator, we learn, is a bearded man standing next to a fire in what looks like a log cabin, while outside, a winter storm blankets the town. This is Salla, Finland, located at 66 degrees north. The coldest place in a damn cold country.

It felt like a preview for a scary movie, the kind that starts with images of normal people living seemingly normal lives in a small town, until things go sideways. In this version, the locals are suntanning in shorts and T-shirts in the snow. A middle-aged man in a blue Speedo enters a frozen lake. And then, Salla's mayor comes into the frame to announce the city wants to host the 2032 Summer Olympic Games. At first, I thought it was a joke. As the video progressed, it became clear that it was not funny at all.

The viewer is taken on a tour of the wintry village, with athletes pointing out where the future locations of outdoor venues would be held: a snowy field would become the beach volleyball court, a forested area would become the mountain bike track, a frozen lake would become the paddling venue, and so on. And then they drop the tagline: "Warm hearts? We have it. Warm place? Coming soon." I was shaken.

The Salla 2032 Olympic parody bid is a climate awareness campaign with a simple premise: if climate change progresses, even the coldest, most northern parts of Finland could become summer destinations. It made the rounds on social media, raising eyebrows and stirring up some climate angst among sports fans. Employing a quintessentially Finnish sense of humour and some clever video work, the campaign shows the world exactly what's at risk, tactfully inspiring action.

A summer-ready Salla is not a foregone conclusion. Neither are any of the more dire predictions I've offered throughout this book. We can choose a different path, adopt an aggressive adaptation plan, protect those most at risk from the worst outcomes, and preserve a healthy future. The key word is "choose".

Sport 2050

In October 2020, I attended the first Sport Positive Summit online and met BBC Producer Dave Lockwood (now Head of Editorial Sustainability at BBC Sport) to chat about climate change coverage in sports news. We spent the first 20 minutes lamenting the lack of climate coverage in sport, and discussing how to insert gloomy climate narratives into the colour commentary during a football match. And then he surprised me. Just before the clock ran out on our call, Dave pitched a project to me that would forecast what sports might look like in 2050 if we do absolutely nothing on climate. A worst-case scenario projection, a bit of harmless scare-mongering. Maybe, he mused, we can inspire people to do something.

Yes, I remember thinking, *this* was a good idea. A little climate anger could stir up some urgency, and frankly too little had been done at that point to inspire action among those furthest from the frontlines and fence lines.

A month later, Dave had assembled a team of researchers to come up with possible storylines for sport in 2050, and to provide evidence for the projection. The committee included

Dr. Nick Watanabe at the University of South Carolina, Dr. Russell Seymour, CEO of the British Association for Sustainability in Sport, a PhD student named Kate Sambrook, Dr. David Goldblatt, a sports historian and writer, and to my delight, me.

The team got on a call and quickly realized this would be more complicated than we had initially thought. Not because we couldn't come up with ideas of what would go wrong, but because we had to assume sports would do some adapting between now and 2050, and that had to be accounted for. The conversations were all over the place. We discussed every region of the globe and every major event series. We talked through worst-case scenarios, and slightly softer versions of every story. We debated whether to rule out some countries from future event hosting altogether, and whether there would still be a Winter Olympics. We reviewed flood maps to figure out whether professional football would be viable. We considered whether water sports would be possible without antibacterial wetsuits (that one, in particular, was a sad thought experiment). We came up with more than a dozen possible storylines, and were told that was too much.

Five stories came out of that process, and were published on the BBC Sport website on 17 May, 2021, with the following disclaimer: "The imagined scenarios are *not* predictions; they are creatively imagined and for illustrative purposes only – but are based on both the science and collective thoughts of how sport might adapt." Here's what we imagined for Sport 2050.

Football

We imagined a World Cup in China with a limited number of traveling fans, indoor stadiums, and night games to avoid the summer heat. We considered that perhaps 45-minute halves are too long, given the heat challenges, so we changed the format to 20-minute thirds as in ice hockey, with rolling substitutions to reduce the likelihood of heat illness. Players will wear biometric monitors that will alert coaches and medical staff if they're injured, overheating, or dehydrated. To reduce the carbon footprint of the event, only 24 teams will qualify for China 2050.

This is not an unlikely future outcome. Already, during COVID, China hosted the Beijing 2022 Olympics with no international fans. In the early 2010s, President Xi Jinping made public his "three wishes" – to qualify for, to host, and to win the World Cup. This is now official policy, and Chinese firms have invested heavily in creating a football presence in the country. With summer temperatures in the country steadily rising, it's already seeing some of the most severe heatwaves in the world. Indoor facilities will be the only viable option for safe play in the summer months.

Cricket

Cricket will also move indoors in this hypothetical worst-case future. That is, if cricket is still played at all. In many parts of the world, drought and heat could make it impossible to maintain league play, and participation could shrink as a result. BBC Sport ran a story of a Test cricket match between Australia and England played in a dome in Melbourne. Costing upward of $3 billion (£1.5 billion), we imagined a 100,000-seat dome being built to control the temperatures and reduce days lost to wildfire smoke. Fan experiences were also considered in the story, with mixed responses to the dome. Some hated it, some accepted it, and all acknowledged it was imperfect, a concession to ensure cricket continued.

Golf

For golf, we considered water availability and thought that maybe maintaining greens and fairways won't be possible in some places. Instead, golf could be played in concrete jungles: an extreme version of the game, played between cooling towers, on the grounds of decommissioned fossil-fuel refineries, and in abandoned sports stadia. The ball would roll not across grass but rather across rubble and bits of concrete, which can send it ricocheting in any direction. A much more 3-D version of the game.

Winter sports
Winter sports are facing the most dire outlook as their programming can be disrupted or shut down entirely by climate change. Although, admittedly, we probably didn't go far enough in our imagined story of skiing in 2050. The story we imagined saw mid- and high-altitude resorts, those that are meant to be more insulated from climate change as they have colder weather, facing partial closures and interrupted seasons. Professional skiers expressed frustration with a competition calendar that was full of cancellations and postponements.

Unfortunately, this future came faster than we anticipated. A winter heatwave across Europe in January 2023 forced nearly 50 per cent of ski resorts to close runs. Several ski competitions were canceled. Mikaela Shiffrin had to wait several extra weeks to contend for the "most wins ever" title, eventually earning a new world record – for both men and women – in March 2023. We clearly underestimated how fast climate change is infringing upon winter sport.

The fifth story
The BBC team rounded out the Sport 2050 series with brilliant animations and graphics, including a video detailing how heat illness works and a "good news quiz" on how sports are tackling climate change. Overall, it went down well.

Doomism and hopium

In the two years since those stories aired, as news cycle after news cycle delivered bad news about climate change and I considered writing this book, I started thinking that maybe we had missed an opportunity with the Sport 2050 story by only sharing the bad stuff. Without trying to, we fell into the trap of doomism. In our effort to galvanize people to act, we may have inadvertently scared them away.

Research on climate messaging, emotions, and willingness to act is relatively new and fast-evolving. A study by Dr. Sander van der Linden at Princeton University took a stab at analyzing a variety of factors influencing climate change risk perception and found sadness or frustration over climate change was the single largest predictor of action. A separate meta-analysis by Drs Anne van Valkengoed and Linda Steg at the University of Gothenberg reviewed 106 studies on motivators of climate change adaptation behaviors and found negative feelings like sadness or anger about climate change to be one of the largest predictors, alongside perceived self-efficacy, descriptive norms (the perception that others are doing something, too) and outcome efficacy. Worry and guilt are also powerful and motivating emotions. Increasingly, though, there's a sense that perhaps learning about climate change and being inundated with negative messaging can be traumatizing and cause some people to shut down.

Hope and optimism are alternatives. These can be triggered with good news stories and information about climate solutions. Yet critics of hope-based appeals have pointed out that emphasizing progress in climate change mitigation to create hope may lead to complacency, as people may not see the need for personal action. Pointing to the tech giants in Silicon Valley and suggesting that technology will fix it, or that going to Mars will fix it, or that it won't be so bad and we may not have to fix it at all, can backfire by letting people off the hook. Offering people some hope is important, but we can't serve up "hopium".* The jury is still out on which emotions are most

---

*A term used by journalists like Amy Westervelt and essayist Mary Annaïse Heglar to describe climate journalism which is all good news, no bad news, no balance.

powerful for inspiring sustained climate action (as opposed to a one-off action), but it's probably some combination of positive and negative, in measured doses. We have to tell the whole truth – it's bad, it'll get worse, it's our fault, and it's our responsibility – while also reminding each other that in its time, humanity has accomplished some pretty incredible things and we can do that again.

Sport has a lot of the language and ideology we need to tackle climate change. Those of us who are fans or who loved sport at any point in our lives will know: you win some, you lose some. It's not always easy. You have to roll with the punches. We need a game plan, coaches, a healthy amount of grit, and a heavy dose of resilience. Anything is possible.

Making it real

A year after the Sport 2050 story aired, I attended the third Sport Positive Summit – this time in person, at Wembley Stadium. The event was also the launch of the BBC Green Sports Awards (yet another of Dave Lockwood's ideas to boost the visibility of sustainability in sport). And again, over a coffee on the too-short break between conference sessions, Lockwood and I had a conversation about Sport 2050. This time, our heads were in a different place. Lockwood was launching awards – it's a good news story, and we'd spent the day listening to some of the best and brightest in the business announcing their latest commitments to carbon neutrality, zero waste, and other goals. The tone was different in this conversation, a far cry from the one we'd had two years earlier. We wondered aloud what sport would look like in 2050 if we got *everything right*.

I had recently been through a similar thought experiment while working with the United Nations Environment Programme on the Sports for Nature Report. The different parties involved in the report saw my initial list and had me scale it back, because they wanted it to be a list of things that sports organizations could control themselves without external input. So I dialed back the ambition and came up with this:*

- Cleaner air near sports sites, due to reduced car traffic, more trees, and strong air quality guidelines among sports organizations.
- All-natural and native grasses for sports turf, which could mean safer conditions for play and lower reliance on chemicals for maintenance.
- No plastic pollution left behind where sport is played.
- No natural spaces lost for new sports facilities, meaning all new facilities are built on brownlands, upcycled urban spaces, or existing sports sites.
- New trees and shrubbery planted at every sports stadium to improve air quality, capture carbon, and provide habitat.
- Sports participants living in harmony with wildlife, respecting their boundaries, and creating feeding and resting grounds for animals.
- Strong partnerships between sports organizations and conservation organizations, to grow awareness of the risks of biodiversity loss and rally sports fans around nature.

But this is my book, so I'm going to go make a longer and more ambitious list. What does sport look like in a world where we stop and reverse the climate crisis, address inequities, restore biodiversity, share wealth, improve health, and find peace?

- The main priority will be health and wellbeing. The second priority will be fun. Revenues will be relegated further down the list.

---

*United Nations Environment Programme (UNEP), 2022. *Sports for Nature: Setting a baseline*. Handbook. Nairobi, Kenya.

- There are as many women leaders and coaches running sports as men.
- There are as many opportunities for women and girls to play sports as there are for men and boys.
- Sports opportunities are made accessible to all, regardless of ability, finances, gender identity, or any other factor, be it trait or circumstance.
- You can watch any sport on TV or on the internet, live, maybe even in 3D holograms? I don't really know the tech space well enough to predict this one, but it'll be great.
- No more short-distance flights. More trains (high speed!) and fancy, decked-out electric team buses.
- Professional sport is organized into regional divisions, with tournaments hosted in single locations for cross-regional games in-season. For example, a North-west and North-east division might meet for an All-North tournament, where they play multiple games over five days in one host city, then go back to their own regions. This way, cross-country trips are reduced, and no one-off cross-country travel happens until the post-season.
- New sports facilities are only built in places that can climatically accommodate the given sport. For example, no more golf courses in deserts. No more winter sport venues in warm climates. Instead, money is invested in sports opportunities that make sense in each place: outdoor tennis and basketball courts with shading in places that are hot and dry, cycling lanes in every city, field sports with proper flood protections and drought-resistant grasses, lined with trees for some shade.
- All facilities powered by renewable energy. I envision solar panels over car parks for the electric cars and buses the teams and fans will use, and wind turbines looming above stadiums wherever possible.
- Sponsors will promote green energy, healthy and affordable foods, and campaigns for equity and inclusion. They won't

be trying to sell us more "stuff", because consumption rates will be way down, planned obsolescence will be a thing of the past, and we'll all be finding ways to make our stuff more durable.

- Sports stores will have second-hand sections as big as their new sections.
- Coaches, parents, and athletes will be aware of climate hazards and know what to do in extreme heat, cold, rain, floods, droughts, or other unsafe conditions. Everybody will be safer, injury rates will come down, and long-term damage linked to concussions will be addressed.
- Sports will be more fun, less pressured.
- Sports will also be more flexible, less structured. The games themselves will be the same but, with more free time in our days (due to better labor standards globally and more automation), we'll be able to adopt flexible game schedules and Plan B options.
- Sports venues will double as community centers, meaning that big stadiums that used to *just* host football on Sundays, or an arena that used to *just* hold a basketball game once or twice a week, are now being used throughout the week as a community space. In case of emergency, these venues are used as shelters, resource distributions sites, or clinics. In election season, they become voting centers.

Thanks to all the people working to make sport more inviting, more accessible, and more climate resilient, the sector and all its participants will thrive.

Is that not worth fighting for?

Author's Note

Like any project, this one had environmental impacts.

My research for this book (and the academic work that informed it) took me to Kenya, Australia, New Zealand, Mexico, Puerto Rico, Scotland, France, all around the US and Canada. Wherever possible, I chose ground transport over flights. Still, it was a *lot* of travel. So I calculated and offset the emissions linked to all travel, including my husband's travel when he tagged along (14 tonnes CO_2e), energy use at home (4.1 tonne CO_2e), and my food consumption for the 12 months I spent working on this book (1.1 tonne CO_2e).

Through LivClean, a women-run Canadian offsetting company, I chose to support the Great Bear Forest Carbon Project, an improved forest management program run by the Haida Nation. The Great Bear Forest Carbon Project covers more than 14 million acres in British Columbia and is home to the largest remaining intact coastal temperate rainforest in the world. By protecting this rainforest through legal protections and ongoing restoration work, the project ensures the long-term protection of many species that cannot be found anywhere else on the planet, such as the Kermode bear and the Western Red Cedar.

Offsets aren't perfect, but I'll take imperfect action over perfect inaction.

Acknowledgments

Warm thanks to my agent, Joe Perry, for seeing the potential of this project when it was just a glimmer of an idea, then championing my career from our first conversation. Cheers Joe, we did it!

To everyone who was interviewed for this book, thanks for speaking up about our new climate realities. I know it's not always easy.

To my editors at Bloomsbury, Jim Martin, Sarah Lambert and Catherine Best, my words were pretty average before you skilfully upgraded them. Thanks also to the whole Bloomsbury marketing and PR team – Lizzy Ewer, Rachel Nicholson, and co. – for making sure this book would find its readers.

To Ann Pegoraro, Tiffany Richardson, and Aileen McManamon – your mentorship has been invaluable. I'm lucky to know and work with you all.

Endless thank-yous are owed to the Institute for Sport Business team at Loughborough University London: James Skinner, Andrea Geurin, Holly Collison-Randall, Lauren Burch, Emily Hayday, Dan Read, Eddie Mighten, Jacky Mueller, and Russ Seymour for the support and encouragement as I wrote this book. I've never laughed more in work meetings than with you lot.

To the Kinesiology and Physical Education team at University of Toronto who welcomed me warmly when I was in the final throes of writing this book and have been great collaborators from the jump.

To the members of Book Club: Jessica Murfree, Lauren Lichterman, Laura Stargel, Kristen Fulmer, Kristin Hanczor, Aly Criscuolo, Monica Rowand, Mara Abbott, and Eileen Quigley, you keep me fiery and focused on the long game. You're my blade of grass.

To the Sport Ecology Group team – thanks for believing in sport ecology with me. I'm a better scholar because I've worked with each of you.

To my friends, spread across the continents, who've heard my many crazy ideas and helped me find the good ones, especially in the last couple years as I wrote this book: Katya Moussatova, Emily Skahan, Tanis Smither, Vanessa Hayford, and Megan McSorley. I appreciate you more than you know.

To my family – Mom, Dad, Steph, Delaney, Pascal, Jordan, Karine, Nico, Ari, Sybil, Brandon, Bruce, Dvorah, and Jan. You are the best support system.

And to Daniel Sailofsky, my favorite person. Thank you for everything. Je t'aime.

Bibliography

Chapter 1: Pregame

Carrington, D., May 17, 2019. "Why the Guardian is changing the language it uses about the environment". *Guardian*. https://www.theguardian.com/environment/2019/may/17/why-the-guardian-is-changing-the-language-it-uses-about-the-environment

Griffin, P., 2017. "The Carbon Majors Database: CDP Carbon Majors Report 2017". 14pp.

Heede, R., 2014. "Tracing anthropogenic carbon dioxide and methane emissions to fossil fuel and cement producers, 1854–2010". *Climatic Change* 122, 229–41. https://doi.org/10.1007/s10584-013-0986-y

IPCC Special Report: "Global Warming at 1.5°C". https://www.ipcc.ch/sr15/

Leiserowitz, A., Carman, J., Buttermore, N., et al., 2022. "International Public Opinion on Climate Change, 2022". New Haven, CT:Yale Program on Climate Change Communication and Data for Good at Meta.

Lloyd's Register Foundation, 2019. *2019 World Risk Poll.* Accessed June 2021. https://wrp.lrfoundation.org.uk/2019-world-risk-poll/

Mann, M.E., 2021. "Beyond the hockey stick: Climate lessons from the Common Era". *Proceedings of the National Academy of Sciences*, 118(39), p.e2112797118.

Orr, M. and Inoue, Y., 2019. "Sport versus climate: Introducing the climate vulnerability of sport organizations framework". *Sport Management Review*, 22(4), pp. 452–63.

Solnit, R., January 12, 2023. "'If you win the popular imagination, you change the game': why we need new stories on climate". *Guardian*. https://www.theguardian.com/news/2023/jan/12/rebecca-solnit-climate-crisis-popular-imagination-why-we-need-new-stories

United Nations, n.d. "Global Issues: Climate Change". Accessed June 2022. https://www.un.org/en/global-issues/climate-change

Chapter 2: Heat Check

Garcia, C.K., Renteria, L.I., Leite-Santos, G., et al., 2022. "Exertional heat stroke: pathophysiology and risk factors". *BMJ Medicine*, 1:e000239. doi: 10.1136/bmjmed-2022-000239

Hayhoe, K., 2022. "Heat". In G. Thunberg (ed.), *The Climate Book*, 1st edn, Allen Lane, pp. 50–2.

Miller, K.C., Casa, D.J., Adams, W.M., et al., 2021. "Roundtable on Preseason Heat Safety in Secondary School Athletics: Prehospital Care of Patients With Exertional Heat Stroke". *Journal of Athletic Training*, 56(4): 372–82. doi: 10.4085/1062-6050-0173.20

Nichols, A.W., 2014. "Heat-related illness in sports and exercise". *Current Reviews in Musculoskeletal Medicine*, 7(4): 355–65. doi: 10.1007/s12178-014-9240-0.

Smith, K. R., Woodward, A., Lemke, B., et al., 2016. "The last Summer Olympics? Climate change, health, and work outdoors". *Lancet*, 388(10045), 642–4. doi:10.1016/S01406736(16)31335-6

Thunberg, G., 2022. "The world has a fever". In G. Thunberg (ed.), *The Climate Book*, 1st edn, Allen Lane, pp. 32–3.

Walters, Inc., Consultant in Sport Medicine, September 21, 2018. "An Independent Evaluation of Procedures and Protocols Related to the June 2018 Death of a University of Maryland Football Student-athlete". Accessed June 2021. https://www.usmd.edu/newsroom/Walters-Report-to-USM-Board-of-Regents.pdf

Chapter 3: Wild Wild(fire) West

Congressional Research Service, 2023. "Wildfire Statistics: Report". Accessed April 2022. https://sgp.fas.org/crs/misc/IF10244.pdf

Gergis, J., 2022. "Wildfires". In G. Thunberg (ed.), *The Climate Book*, 1st edn, Allen Lane, pp. 96–8.

Grennan, G. K., Withers, M. C., Ramanathan, D. S., and Mishra, J., 2023. "Differences in interference processing and frontal brain function with climate trauma from California's deadliest wildfire". *Plos Climate*, 2(1), e0000125.

Isaac, F., Toukhsati, S.R., Di Benedetto, M. and Kennedy, G.A., 2021. "A Systematic Review of the Impact of Wildfires on Sleep Disturbances". *International Journal of Environmental Research and Public Health*, 18(19), p. 10,152.

Jones, M.W., Abatzoglou, J.T., Veraverbeke, S., et al., 2022. "Global and regional trends and drivers of fire under climate change". *Reviews of Geophysics*, 60(3), p.e2020RG000726.

Romps, D.M., Seeley, J.T., Vollaro, D. and Molinari, J., 2014. "Projected increase in lightning strikes in the United States due to global warming". *Science*, 346(6211), pp. 851–4.

Short, Karen C., 2017. *Spatial wildfire occurrence data for the United States, 1992–2015* [FPA_FOD_20170508], 4th edn. Fort Collins, CO: Forest Service Research Data Archive. https://doi.org/10.2737/RDS-2013-0009.4

Silveira, S., Kornbluh, M., Withers, M.C., et al., 2021. "Chronic mental health sequelae of climate change extremes: A case study of the deadliest Californian wildfire". *International Journal of Environmental Research and Public Health*, 18(4), p. 1,487.

United Nations Environment Programme, 2022. "Spreading like Wildfire – The Rising Threat of Extraordinary Landscape Fires". UNEP Rapid Response Assessment, Nairobi.

Chapter 4: Every Breath You Take

American Forests, 2021. "American Forests Launches Nationwide Tree Equity Scores". Accessed June 2022. https://www.americanforests.org/article/american-forests-launches-nationwide-tree-equity-scores/

Archsmith, J., Heyes, A. and Saberian, S., 2018. "Air quality and error quantity: Pollution and performance in a high-skilled, quality-focused occupation". *Journal of the Association of Environmental and Resource Economists*, 5(4), pp. 827–63.

Casper, J.M. and Bunds, K.S., 2018. "Tailgating and air quality". In Brian P. McCullough and Timothy B. Kellison (eds), *Routledge Handbook of Sport and the Environment*, Routledge, pp. 291–300.

Chen, Y., Jin, G.Z., Kumar, N. and Shi, G., 2011. "The promise of Beijing: Evaluating the impact of the 2008 Olympic Games on air quality". NBER (National Bureau of Economic Research) Working Paper No. 16907.

Cusick, M., Rowland, S.T. and DeFelice, N., 2023. "Impact of air pollution on running performance". *Scientific Reports*, 13(1), p. 1,832.

Glazener, A., Wylie, J., van Waas, W., et al., 2022. "The Impacts of Car-Free Days and Events on the Environment and Human Health". *Current Environmental Health Reports*, 9, pp. 165–82.

Lapere, R., Menut, L., Mailler, S. and Huneeus, N., 2020. "Soccer games and record-breaking PM 2.5 pollution events in Santiago, Chile". *Atmospheric Chemistry and Physics*, 20(8), pp. 4,681–94.

Li, Y., Wang, W., Kan, H., Xu, X. and Chen, B., 2010. "Air quality and outpatient visits for asthma in adults during the 2008 Summer Olympic Games in Beijing". *Science of the Total Environment*, 408(5), pp. 1,226–7.

London Air, n.d. "Air Pollution Research in London". Accessed June 2022. https://www.londonair.org.uk/LondonAir/General/research. aspx#:~:text=Air%20pollution%20is%20a%20very,acts%20as%20a%20 giant%20laboratory

Nowak, D.J., Hirabayashi, S., Bodine, A. and Greenfield, E., 2014. "Tree and forest effects on air quality and human health in the United States". *Environmental Pollution*, 193, pp. 119–29.

World Health Organization, 2021. "WHO global air quality guidelines: particulate matter (PM2.5 and PM10), ozone, nitrogen dioxide and carbon monoxide". 290pp. https://www.who.int/publications/i/ item/9789240034228

—, 2022. "Ambient (outdoor) air pollution". Accessed June 2022. https:// www.who.int/news-room/fact-sheets/detail/ambient-(outdoor)-air- quality-and-health

Yang, B., Liu, H., Kang, E.L., et al., 2022. "Traffic restrictions during the 2008 Olympic Games reduced urban heat intensity and extent in Beijing". *Communications Earth & Environment*, 3(1), p. 105.

Chapter 5: Playing on the Edge

Oppenheimer, M., Glavovic, B.C., Hinkel, J., et al., 2019. "Sea Level Rise and Implications for Low-Lying Islands, Coasts and Communities". In *IPCC Special Report on the Ocean and Cryosphere in a Changing Climate*, H.-O. Pörtner, D.C. Roberts, V. Masson-Delmotte, et al. (eds), Cambridge University Press, pp. 321–445.

Reineman, D.R., Thomas, L.N. and Caldwell, M.R., 2017. "Using local knowledge to project sea level rise impacts on wave resources in California". *Ocean & Coastal Management*, 138, pp. 181–91.

Rennie, A.F., Hansom, J.D., Hurst, M.D., et al., 2021. "Dynamic Coast: Adaptation and Resilience Options for St Andrews Links", CRW2017_08. Scotland's Centre of Expertise for Waters (CREW). Available online at: crew.ac.uk/publications

Sports and Fitness Industry Association, 2022. "2022 Surfing Participation
    Report". Accessed January 2023.
Vousdoukas, M.I., Ranasinghe, R., Mentaschi, L., et al., 2020. "Sandy coast-
    lines under threat of erosion". *Nature Climate Change*, 10(3), pp. 260–3.

Chapter 6: Come Hell or High Water

Busfield, S., October 12, 2019. "Typhoon stops play at Rugby
    World Cup". *Forbes* magazine. https://www.forbes.com/sites/
    stevenbusfield/2019/10/12/typhoon-hagibis-stops-play-at-rugby-
    world-cup/?sh=750b0ad73513
Finney, B., 2021. "Conservation: Healing Waters". *Oceanographic Magazine*.
    https://oceanographicmagazine.com/features/planet-patrol/
Goldblatt, D., 2020. "Playing against the clock: Global sport, the climate
    emergency and the case for rapid change". Rapid Transition Alliance.
    https://rapidtransition.org/resources/playing-against-the-clock/
IPCC, 2022. "Climate Change 2022: Impacts, Adaptation, and
    Vulnerability". Contribution of Working Group II to the *Sixth
    Assessment Report of the Intergovernmental Panel on Climate Change*, H.-O.
    Pörtner, D.C. Roberts, M. Tignor, et al. (eds). Cambridge University
    Press, doi:10.1017/9781009325844
Kummu, M., De Moel, H., Ward, P.J. and Varis, O., 2011. "How close do
    we live to water? A global analysis of population distance to freshwater
    bodies". *PloS One*, 6(6), p.e20578.
Li, S. and Otto, F.E., 2022. "The role of human-induced climate change
    in heavy rainfall events such as the one associated with Typhoon
    Hagibis". *Climatic Change*, 172(1–2), p. 7.
Rahmstorf, S., 2022. "Warming oceans and rising seas". In G. Thunberg
    (ed.), *The Climate Book*, 1st edn, Allen Lane, pp. 78–83.
Tanhua, T., Gutekunst, S.B. and Biastoch, A., 2020. "A near-synoptic
    survey of ocean microplastic concentration along an around-the-world
    sailing race". *Plos One*, 15(12), p.e0243203.

Chapter 7: Dust Bowl

Behere, P.B., Chowdhury, D., Behere, A.P., and Yadav, R., 2021. "Psychosocial
    aspects of suicide in largest industry of farmers in Vidarbha Region of
    Maharashtra". *Industrial Psychiatry Journal*, 30(Suppl. 1), S10.
Chiew, F. and Prosser, I., 2011. "Water and climate". In Prosser, I. (ed.),
    Science and solutions for Australia, CSIRO Publishing, pp. 29–46.

Manchester City FC, 2020. "Manchester City and Xylem present: The
    end of football". Video, available online: https://www.mancity.com/
    features/end-of-football/

Milly, P.C. and Dunne, K.A., 2020. "Colorado River flow dwindles as
    warming-driven loss of reflective snow energizes evaporation". *Science*,
    367(6483), pp. 1,252–5.

Seifert, K., April 20, 2020. "NFLPA: New injury data shows grass
    'significantly safer' than turf". ESPN. Accessed March 2023. https://
    www.espn.co.uk/nfl/story/_/id/36243906/nflpa-new-injury-data-
    shows-grass-significantly-safer-turf

Tretter, J., 2020. "Only natural grass can level the NFL's playing field".
    NFLPA (National Football League Players Association). Accessed July
    2022. https://nflpa.com/posts/only-natural-grass-can-level-the-nfls-
    playing-field

—, 2023. "Why the NFL's approach to field surfaces is uneven". NFLPA.
    Accessed March 2023. https://www.espn.co.uk/nfl/story/_/
    id/36243906/nflpa-new-injury-data-shows-grass-significantly-safer-turf

UN Water, 2018. "Water scarcity report". Accessed March 2023. https://
    www.unwater.org/sites/default/files/app/uploads/2018/10/
    WaterFacts_water-scarcity_sep2018.pdf

Vins, H., Bell, J., Saha, S. and Hess, J.J., 2015. "The mental health outcomes
    of drought: a systematic review and causal process diagram". *International
    Journal of Environmental Research and Public Health*, 12(10), pp. 13,251–75.

Chapter 8: Shelter of Last Resort

Adelson, J., May 22, 2019. "Hurricane season changes for New Orleans:
    Smoothie King Center to be evacuation staging area". https://www.
    nola.com/hurricane-season-changes-for-new-orleans-smoothie-king-
    center-to-be-evacuation-staging-area/article_28c57089-8d19-5234-
    8d2b-eee8c9be0aeb.html

Association of British Insurers, 2005. "Financial Risks of Climate
    Change". https://insurance.lbl.gov/documents/abi-climate.pdf

BBC, August 15, 2015. "Katrina anniversary: Inside the Superdome during
    Katrina". https://www.bbc.co.uk/news/av/magazine-33943911

Guarino, M., August 15, 2015. "Katrina firsthand: Superdome general
    manager recounts chaos in days after Katrina". *The Advocate*. Accessed
    June 2021. https://www.theadvocate.com/baton_rouge/news/katrina-
    firsthand-superdome-general-manager-recounts-chaos-in-days-after-
    katrina/article_dad814cc-7daf-557d-8455-5b6e29f0f4cc.html

Holland, G. and Bruyère, C.L., 2014. "Recent intense hurricane response to global climate change". *Climate Dynamics*, 42, pp. 617–27.

Kellison, T., Orr, M. and Watanabe, N.M., 2023. "The nonexcludable function of sports stadiums in climate-changed cities". *Sport in Society*, pp. 1–20.

NOAA (National Oceanic and Atmospheric Administration), n.d. "Hurricane costs". Accessed August 2022. https://coast.noaa.gov/states/fast-facts/hurricane-costs.html

Spong, John, 2005. "Dome Away From Home". *Texas Monthly*, 154–9, lines 261–8.

Chapter 9: Let It Snow

Albrecht, G., 2006. "Solastalgia". *Alternatives Journal*, 32(4/5), pp. 34–6.

Climate Champions Podcast, 2020. Season 3, Episode 7: Henri Rivers on skiing. Sport Ecology Group.

Coleman, A.G., 1996. "The unbearable whiteness of skiing". *Pacific Historical Review*, 65(4), pp. 583–614.

de Jong, C., 2011. "Artificial Production of Snow". In V.P. Singh, P. Singh and U.K. Haritashya (eds), *Encyclopedia of Snow, Ice and Glaciers*, Springer, pp. 61–6.

FIS, 2023. "FIS welcomes athlete engagement on sustainability". Accessed March 2023. https://www.fis-ski.com/en/international-ski-federation/news-multimedia/news-2022/fis-welcomes-athlete-engagement-on-sustainability

Klein, G., Vitasse, Y., Rixen, C., Marty, C. and Rebetez, M., 2016. "Shorter snow cover duration since 1970 in the Swiss Alps due to earlier snowmelt more than to later snow onset". *Climatic Change*, 139, pp. 637–49.

Lynch, F.L., Peterson, E.L., Lu, C.Y., et al., 2020. "Substance use disorders and risk of suicide in a general US population: a case control study". *Addiction Science & Clinical Practice*, 15(1), pp. 1–9.

Protect Our Winters, 2023. FIS athlete letter to demand climate action. Accessed March 2023. https://protectourwinters.eu/wp-content/uploads/2023/03/open-letter-to-FIS-230209.pdf

Riley, H., 2016. "Endlings". *Island Magazine*, 146. https://islandmag.com/read/endlings-by-harriet-riley

Robinson, David A., Estilow, Thomas W., and NOAA CDR Program, 2012. "NOAA Climate Data Record (CDR) of Northern Hemisphere (NH) Snow Cover Extent (SCE)", Version 1. NOAA National Centers for Environmental Information. Accessed October 2021.

Samuel, H., January 6, 2023. "Elite Swiss ski resort flies snow to the slopes after mild winter". *Daily Telegraph*. https://www.telegraph.co.uk/world-news/2023/01/06/elite-swiss-ski-resort-flies-snow-slopes-mild-winter/

Scott, D., Knowles, N.L., Ma, S., Rutty, M. and Steiger, R., 2023. "Climate change and the future of the Olympic Winter Games: athlete and coach perspectives". *Current Issues in Tourism*, 26(3), pp. 480–95.

University of Waterloo, 2023. "Climate change threatens future Winter Olympics". https://uwaterloo.ca/news/media/climate-change-threatens-future-winter-olympics

Zeng, X., Broxton, P. and Dawson, N., 2018. "Snowpack change from 1982 to 2016 over conterminous United States". *Geophysical Research Letters*, 45(23), pp. 12,940–7.

Chapter 10: Thin Ice

Agence France-Presse, February 15, 2023. "World's largest skating rink on thin ice as Canada's winter prevents opening". *Guardian*. https://www.theguardian.com/world/2023/feb/15/rideau-canal-skateway-worlds-largest-skating-rink-ottowa-canada-warm-winter

Angé, O. and Berliner, D., eds, 2020. *Ecological nostalgias: Memory, affect and creativity in times of ecological upheavals*, vol. 26, *Environmental Anthropology and Ethnobiology*, Berghahn Books.

Brammer, J.R., Samson, J., & Humphries, M.M., 2015. "Declining availability of outdoor skating in Canada". *Nature Climate Change*, 5(1), pp. 2–4.

Futterman, M., 2020. "Sledding athletes are taking their lives. Did brain-rattling rides and high speed crashes damage their brains?" *New York Times*, available online. https://www.nytimes.com/2020/07/26/sports/Olympics/Olympics-bobsled-suicide-brain-injuries.html

Hoogbergen, A.V.F., 2021. *Waiting for Yesteryears: An ethnographic exploration of ecological nostalgia in the contemporary Dutch intangible cultural heritage context*, Master's thesis, Utrecht University.

Koolhaas, Marnix, 2010. *Schaatsenrijden: Een cultuurgeschiedenis*. Amsterdam, Veen.

McCradden, M.D. and Cusimano, M.D., 2018. "Concussions in sledding sports and the unrecognized 'sled head': a systematic review". *Frontiers in Neurology*, 9, p. 772.

RinkWatch, 2022. "2021–2022 RinkWatch Report". https://www.rinkwatch.org/documents/rinkwatch_report_2021-2022.pdf

Robertson, C., McLeman, R. and Lawrence, H., 2015. "Winters too
    warm to skate? Citizen-science reported variability in availability of
    outdoor skating in Canada". *The Canadian Geographer/Le Géographe
    canadien*, 59(4), pp. 383–90.
Sharma, S., Blagrave, K., Magnuson, J.J., et al., 2019. "Widespread
    loss of lake ice around the Northern Hemisphere in a warming
    world". *Nature Climate Change*, 9(3), pp. 227–31.
Sivanathan, S., Oleon, A., Thanapalan, K.K.T. and McCarthy, P., 2019.
    "Real-Time Wireless Vibration Logger Skeleton Run", 2019 25th
    International Conference on Automation and Computing (ICAC),
    Lancaster, UK, pp. 1–5. doi: 10.23919/IConAC.2019.8895130
Visser, H. and Petersen, A.C., 2009. "The likelihood of holding outdoor
    skating marathons in the Netherlands as a policy-relevant indicator of
    climate change". *Climatic Change*, 93, pp. 39–54.

Chapter 11: At a Disadvantage

Butler, L. and Levey, S., 2022. "Climate change likely increased heavy
    rain that led to deadly floods in Pakistan". Imperial College London.
    https://www.imperial.ac.uk/news/239974/climate-change-likely-
    increased-heavy-rain
Evans, S., 2021. "Analysis: Which countries are historically responsible for
    climate change?" *Carbon Brief*. https://www.carbonbrief.org/analysis-
    which-countries-are-historically-responsible-for-climate-change/
Griffin, P., 2017. "The Carbon Majors Database: CDP Carbon Majors
    Report 2017". 14pp. https://cdn.cdp.net/cdp-production/cms/reports/
    documents/000/002/327/original/Carbon-Majors-Report-2017.pdf
Ritchie, H., Roser, M. and Rosado, P., 2020. "CO_2 and Greenhouse Gas
    Emissions". https://ourworldindata.org/co2-and-greenhouse-gas-
    emissions
Sport Positive Summit, 2022. "The challenge of net zero and scope 3 for
    sport". Recorded panel.
The Iten Municipality, 2019. "Integrated Development Plan 2019–2023".
    https://elgeyomarakwet.go.ke/?mdocs-file=2267

Chapter 12: Playing Catch-up

International Olympic Committee, December 6, 2022. "Future Host
    Commission studying landscape of winter sport with a view to the

Olympic Winter Games 2030 and beyond". https://olympics.com/
ioc/news/future-host-commission-studying-landscape-of-winter-
sport-with-a-view-to-the-olympic-winter-games-2030-and-beyond
—, 2022. "Future Host Questionnaire". https://stillmed.
olympics.com/media/Documents/Olympic-Games/Brisbane-2032/
General/Future-Host-Questionnaire-Olympic-Games.pdf

Chapter 13: Back Story

Global Web Index, 2019. "Sports around the world: Examining how
digital consumers around the world engage with sports, and the
opportunities for sports brands and marketers". https://www.gwi.
com/reports/sports-around-the-world
—, 2021. "The Sports Playbook". https://www.gwi.com/reports/the-
sports-playbook
IUCN, 2022. "Sports for Nature". https://www.iucn.org/sites/default/
files/2023-08/sports-for-nature-info-pack.pdf
LifeTackle, 2020. "The environmental awareness and behaviour of
professional football supporters: an empirical survey". https://lifetackle.
eu/assets/files/LIFE_TACKLE_Report_on_supporters_survey.pdf
Lund, M., 1971. "Their new Alps, our new Alps: The coming despoliation
of Colorado". *Ski Magazine*, December 1971.
Müller, M., Wolfe, S.D., Gaffney, C., et al., 2021. "An evaluation of the
sustainability of the Olympic Games". *Nature Sustainability*, 4(4),
pp. 340–8.
UNFCCC, 2018. "Sports for Climate Action", version 2.0. https://
unfccc.int/sites/default/files/resource/Sports_for_Climate_Action_
Declaration_and_Framework.pdf

Chapter 14: Green Sports

GreenSportsPod, 2020. Episode 1, Guest: Dr. Allen Hershkowitz.
International Olympic Committee, 2022. "IF sustainability project:
Carbon fiber circular alliance". https://stillmed.olympics.com/media/
Documents/Beyond-the-Games/Sustainability/IF-Sustainability-Case-
Studies/Carbon-Fibre-Alliance.pdf
Johnson, A.E., 2022. "How to find joy in climate action". TED Talk.
https://www.ted.com/talks/ayana_elizabeth_johnson_how_to_find_
joy_in_climate_action?language=en

New York Road Runners, 2019. "NYRR Sustainability". https://www. nyrr.org/community/nyrr-sustainability

Orr, M., Murfree, J. and Seymour, R., 2023. "Sustainability managers in sport: A new role (ill)defined". Presentation at European Association for Sport Management Conference, September 2023, Belfast, Northern Ireland.

Perez, J., 2021. "SBJ Dealmakers: Leiweke implores industry to fight global warming". *Sports Business Journal.* https://www.sportsbusinessjournal. com/Daily/Issues/2021/12/01/Facilities/Dealmakers-OVG.aspx

Sport Positive, 2023. "Leagues". Accessed April 25, 2023. www. sportpositiveleagues.com

Waste Management Phoenix Open, 2022. "Sustainability Report". https:// www.wm.com/content/dam/wm/assets/inside-wm/phoenix- open/2022-WM-Phoenix-Open-Sustainability-Report.pdf

Wynes, S., 2021. "COVID-19 disruption demonstrates win-win climate solutions for major league sports". *Environmental Science & Technology,* 55(23), pp. 15,609–15.

Chapter 15: In Bed with Big Oil

The American Birkebeiner, 2021. "To the Birkie community". Accessed January 2023. https://www.birkie.com/we-hear-you/

Australian Conservation Foundation, 2022. "Out of bounds: Coal, gas, and oil sponsorship in Australian sports". https://www.acf.org.au/ new-research-fossil-fuel-industry-using-sport-to-greenwash-public- image

Badvertising, 2022. "British Cycling, tell Shell 'on yer bike!'" Accessed January 2023. https://www.badverts.org/latest/british-cycling-tell- shell-on-yer-bike

Brand Essence Market Research, 2022. "Sports sponsorship market". Accessed January 2023. https://brandessenceresearch.com/technology- and-media/sports-sponsorship-market-size

Global Justice Now, 2017. "Corporations data 2017". Accessed January 2023. https://www.globaljustice.org.uk/news/69-richest-100-entities- planet-are-corporations-not-governments-figures-show/

Gough, C., 2021. "Sports sponsorship – Statistics & Facts". *Statista.* Accessed October 2022. https://www.statista.com/topics/1382/sports-sponsorship/

INEOS, n.d. Website home page. Accessed January 2023. ineos.com

Nielsen Sports, 2022. "2022 Nielsen sports report". Accessed January 2023. https://www.nielsen.com/insights/2022/fans-are-changing-the-game/

Rapid Transition Alliance, 2021. "Sweat not oil: Why sports should drop advertising and sponsorship from high-carbon polluters". https:// rapidtransition.org/resources/sweat-not-oil-why-sports-should-drop-advertising-and-sponsorship-from-high-carbon-polluters

Stopping Tobacco Organizations and Products (STOP), 2020. "Driving addiction: F1 and tobacco advertising". https://exposetobacco.org/campaigns/driving-addiction/

Supran, G. and Oreskes, N., 2017. "Assessing ExxonMobil's climate change communications (1977–2014)". *Environmental Research Letters*, 12(8), 084019.

Union of Concerned Scientists, 2015. "The climate deception dossiers: Internal fossil fuel industry memos reveal decades of corporate disinformation". https://www.ucsusa.org/sites/default/files/attach/2015/07/The-Climate-Deception-Dossiers.pdf

WHO, 2023. "Fact sheet: Tobacco". https://www.who.int/news-room/fact-sheets/detail/tobacco

Chapter 16: The Roster: Athletes vs Climate Change

Haugseth, J.F. and Smeplass, E., 2022. "The Greta Thunberg effect: A study of Norwegian Youth's Reflexivity on Climate Change". *Sociology*, 00380385221122416.

Henderson, J., January 20, 2023. "Eco-friendly Innes Fitzgerald turns down World Cross trip". *Athletics Weekly*. https://athleticsweekly.com/athletics-news/eco-friendly-innes-fitzgerald-turns-down-world-cross-trip-1039964384/

International Biathlon Union, 2022. "Biathletes worry about impact of climate change". Accessed August 2022. https://www.biathlonworld.com/news/athlete-sustainability-survey-2021/7iYgxK53MQArnu4JXQfpQN

Maese, R. and Guskin, E., September 10, 2010. "Most Americans support athletes speaking out, say anthem protests are appropriate, poll finds". *Washington Post*. Accessed January 2021. https://www.washingtonpost.com/sports/2020/09/10/poll-nfl-anthem-protests/

MIT Media Lab, 2019. "Project: Pantheon". Accessed January 2021. https://www.media.mit.edu/projects/pantheon-new/overview/

Monbiot, G., August 6, 2008. "I'd rather be a hypocrite than a cynic like Julie Burchill". *Guardian*. https://www.theguardian.com/commentisfree/2008/aug/06/activists.kingsnorthclimatecamp

Sabherwal, A., Ballew, M.T., van Der Linden, S., et al., 2021. "The Greta Thunberg Effect: Familiarity with Greta Thunberg predicts intentions to engage in climate activism in the United States". *Journal of Applied Social Psychology*, 51(4), pp. 321–33.

UK Sport, 2023. "New research shows British public support athletes speaking out and driving social change". Accessed October 2023. https://www.uksport.gov.uk/news/2023/03/08/new-research-shows-british-public-support-athletes-speaking-out-and-driving-social-change

World Athletics, 2021. "World Athletics survey: nearly 80% of athletes seriously concerned about the climate crisis". Accessed August 2022. https://worldathletics.org/athletics-better-world/news/athlete-survey-cop26-united-nations-climate-summit

Chapter 17: What's Next?

BBC Sport, 2021. "Sport 2050: Why are we doing this – and why does it matter today?". Accessed May 2021. https://www.bbc.co.uk/sport/56972366

Van der Linden, S., 2017. "Determinants and measurement of climate change risk perception, worry, and concern". In Matthew C. Nisbet, ed., *The Oxford Encyclopedia of Climate Change Communication*, Oxford University Press.

van Valkengoed, A.M. and Steg, L., 2019. "Meta-analyses of factors motivating climate change adaptation behaviour". *Nature Climate Change*, 9(2), pp. 158–63.

Further reading and viewing

Most of what I know about climate change and sports, I learned through academic research, field work, and conversations with industry experts. But there's so much more to this story than what one book will allow, and there are so many incredible people doing the work to "green" the sports sector. So I've collated a list of some of my favorite sources of information and insight for you to check out.

Books

- *All We Can Save*, Ayanna Elizabeth Johnson and Katherine K. Wilkinson (eds)
- *Clearing the Air*, Tim Smedley

- *The Climate Book*, Greta Thunberg (ed.)
- *Climate Change is Racist*, Jeremy Williams
- *The Future Earth*, Erik Holthaus
- *The Ministry for the Future*, Kim Stanley Robinson (OK, this one is fiction, but it's got so many good and hopeful ideas for how we can improve our future outlooks)
- *Our Biggest Experiment*, Alice Bell
- *Rising: Dispatches from the New American Shore*, Elizabeth Rush
- *We Can't Run Away From This: Racing to Improve Running's Footprint in Our Climate Emergency*, Damian Hall

Podcasts

- Climate Champions (yes, this is a shameless plug for my own podcast, don't @ me)
- Emergency on Planet Sport, hosted by Jonathan Overend (and Melissa Wilson in season 2)
- Green Sports Pod, hosted by Lew Blaustein
- The Sustainability Report, hosted by Matthew Campelli
- Sustaining Sport, hosted by Ben Mole

Documentaries

- *Cricket's Climate Crisis* by Sky Sports (free on YouTube)
- *Football's Toughest Opponent* by Sky Sports (free on YouTube)
- *Jumbo Wild*, a documentary about how British Columbia's remote and sacred Jumbo Valley was *almost* turned into a ski resort
- *Saving Snow*, a 52-minute documentary of winter sports on life support

Organizations dedicated to the cause

- Acting Green Forum (actinggreenforum.com)
- Athletes of the World (athletesoftheworld.org)
- Australian Sports Climate (australiansportsclimate.com.au)
- British Association for Sustainability in Sport (basis.org.uk)
- Council for Responsible Sport (councilforresponsiblesport.org)
- EcoAthletes (ecoathletes.org)
- Football for Future (footballforfuture.org)
- Fossil-Free Football (fossilfreefootball.org)
- Frontrunners (frontrunners.org.au)

- GEO Foundation (sustainable.golf)
- Green Sports Alliance (greensportsalliance.org)
- Pledgeball (pledgeball.org)
- Protect Our Winters (protectourwinters.org)
- Save Pond Hockey (savepondhockey.org)
- Sport and Sustainability International (sportsustainability.org)
- Sport Ecology Group (sportecology.org)
- Sport Positive Summit (sportpositive.org)
- Sports Environment Alliance (sportsenvironmentalliance.org)
- Surfers Against Sewage (sas.org.uk)
- Surfrider Foundation (surfrider.org)

… and so, so many more.

Climate action is a team sport. Welcome to the team.

Index